Extended Performance Evaluation Based on DEA

European University Studies

Europäische Hochschulschriften

Publications Universitaires Européennes

Series V	**Economics and Management**
Reihe V	Volks- und Betriebswirtschaft
Série V	Sciences économiques

Volume/Band **3446**

Ludmila Neumann

Extended Performance Evaluation Based on DEA

A Multidimensional Point of View

PL ACADEMIC RESEARCH

Bibliographic Information published by the Deutsche Nationalbibliothek
The Deutsche Nationalbibliothek lists this publication in the Deutsche Nationalbibliografie; detailed bibliographic data is available in the internet at http://dnb.d-nb.de.

Zugl.: Braunschweig, Techn. Univ., Diss., 2015

D 84
ISSN 0721-7339
ISBN 978-3-631-72216-9 (Print)
E-ISBN 978-3-631-72270-1 (E-PDF)
E-ISBN 978-3-631-72271-8 (EPUB)
E-ISBN 978-3-631-72272-5 (MOBI)
DOI 10.3726/b11297

PL Academic Research is an Imprint of Peter Lang GmbH.
Peter Lang – Frankfurt am Main · Bern · Bruxelles · New York · Oxford · Warszawa · Wien

This publication has been peer reviewed.

www.peterlang.com

Acknowledgments

Writing this dissertation has been certainly the most challenging experience for me and it would not have been possible without the support and guidance that I have received from many people. I would like to take the opportunity to acknowledge and thank all those people who have contributed in many ways to the success of this work.

I would like first of all to express my gratitude to my supervisor, Prof. Dr. Heinz Ahn, for the confidence he has placed in me and for his advice, interest, and encouragement in my research, which were invaluable for the completion of this dissertation. I would also like to thank Prof. Dr. Susanne Robra-Bissantz and Prof. Dr. Thomas S. Spengler for serving as members of my doctoral dissertation committee.

My time at the Institute of Management Control and Business Accounting of the Technische Universität Braunschweig has been made enjoyable and productive in large part due to the great team I have been fortunate to work with over the years. I am especially grateful to Nadia Vazquez Novoa for our inspiring discussions, for the long evenings we have been working together by stimulating and motivating each other, and for all the fun we have had in and out of work. My special thanks go also to Dr. Yvonne Höfer-Diehl for her enormous support and advice – despite physical distance and time difference – in the final period of my dissertation: Your feedback and comments have been of inestimable value.

In my personal life, I would like to thank my parents, Katarina and Leopold Neumann, and my brother, Alex Neumann, for all their love and incredible belief in me throughout every condition of my life. I am also very thankful to my friends for their patience and understanding for the time we could not spend together while I have been focused on writing my dissertation and for all the moral support they have provided.

Finally, I would like to express my gratitude to my husband, Federico Muñoz, who has been by my side throughout this period, tirelessly listening to my ideas and offering encouragement and motivation when it was most needed: Without you, I would not have had the strength to keep up on this challenging project. To you I dedicate this dissertation.

Table of contents

List of figures

List of tables

List of abbreviations

ADD	Additive model
ADD-V	Additive model with variable returns to scale
AR	Assurance region model
AR-I-C	Input-oriented assurance region model with constant returns to scale
AR-I-V	Input-oriented assurance region model with variable returns to scale
AR-O-C	Output-oriented assurance region model with constant returns to scale
AR-O-V	Output-oriented assurance region model with variable returns to scale
BCC	Banker-Charnes-Cooper model
BCC-I	Input-oriented Banker-Charnes-Cooper model
BCC-O	Output-oriented Banker-Charnes-Cooper model
BoD	Benefit-of-the-Doubt model
BSC	Balanced scorecard
C	Constant
CCR	Charnes-Cooper-Rhodes model
CCR-I	Input-oriented Charnes-Cooper-Rhodes model
CCR-O	Output-oriented Charnes-Cooper-Rhodes model
CED	Cost of expired drugs
CO_2	Carbon dioxide
CR	Cone ratio model
CR-I	Input-oriented cone ratio model
CR-O	Output-oriented cone ratio model
CRS	Constant returns to scale
CSP	Cost of service personnel
CVF	Competing values framework
CWT	Number of complaints about waiting time in the store
DEA	Data envelopment analysis
DMU	Decision making unit
DRS	Decreasing returns to scale
G	Generalized
GDEA	Generalized DEA

GDEA-C	Generalized DEA model with constant returns to scale (also referred to as GCCR)
GDEA-V	Generalized DEA model with variable returns to scale (also referred to as GBCC)
HCS_{norm}	Number of health counselling services in the store (normalized values)
HHC_{norm}	Telephone hotline hours for health counselling services (normalized values)
I	Input-oriented
IRS	Increasing returns to scale
KPI	Key performance indicator
MAX	Maximization of performance objectives
MCDM	Multi-criteria decision making
MIN	Minimization of performance objectives
MODEA	Multi objective DEA model
MPI	Malmquist productivity index
MPI-I-C	Input-oriented Malmquist productivity index with constant returns to scale
MPI-I-V	Input-oriented Malmquist productivity index with variable returns to scale
MPI-O-C	Output-oriented Malmquist productivity index with constant returns to scale
MPI-O-V	Output-oriented Malmquist productivity index with variable returns to scale
O	Output-oriented
OE	Organizational effectiveness
OTC	Number of the purchases of over-the-counter products
PEP	Performance evaluation process
PMP	Performance management process
PMS	Performance measurement system
RAM	Range adjusted measure model
RES	Responses to an annual competition in each store's catchment area
RQ	Research question
SDD	Number of subsequent drug deliveries
SQM	Store square meters
TOTC	Turnover of over-the-counter sales
TURN	Absolute increase of the turnover

V	Variable
VRS	Variable returns to scale
WH	Worked hours
Window	Window analysis
Window-I-C	Input-oriented window analysis with constant returns to scale
Window-I-V	Input-oriented window analysis with variable returns to scale
Window-O-C	Output-oriented window analysis with constant returns to scale
Window-O-V	Output-oriented window analysis with variable returns to scale
WR	Weight restriction

List of symbols

Indices

α	Number of red DMUs
δ	Number of green DMUs
γ	Number of yellow DMUs
g	Index of cost objectives
h	Index of benefit objectives
i,t	Indices of outputs
j	Index of inputs
m	Number of inputs
n	Number of outputs
o	Indicator for DMU under investigation (o for observed)
π	Number of DMUs
ρ	Index of DMUs
r	Number of benefit objectives
s	Number of cost objectives
τ	Index for inputs and outputs
ξ	Number of performance objectives
ζ	Index of performance objectives

Variables

AM	Average DMU
β^0	Balance score of DMU under investigation
$\mathcal{C}$	Acceptable output mix region (a multicone)
$\mathcal{C}_{i,t}$	Intersection of multicone and $y_i y_t$ -plane
$c_{g,\tau}$	Multiplier of undesirable inputs and outputs for cost objective g
$\overline{d}_i$	Average absolute deviation from the mean of output i
dev^ρ	Deviation score of DMU ρ
$e_{h,\tau}$	Multiplier of desirable inputs and outputs for benefit objective h

$\varepsilon_1, \varepsilon_2$	Small positive values
η^0	Optimal value of GBCC-MAX model for DMU under investigation
η^0_*	Optimal value of BCC-O model for DMU under investigation
I	Target (output) vector
K	Weighted sum of cost objectives
k_g^ρ	Value of cost objective g for DMU ρ
λ^ρ	Weight of DMU ρ
L	Weighted sum of benefit objectives
l_h^ρ	Value of benefit objective h for DMU ρ
$low_{i,t}$	Slope of lower boundary of intersection $C_{i,t}$ of multicone and $y_i y_t$-plane
μ_h	Weight or multiplier of benefit objective h
$\tilde{\mu}_\tau$	Benefit multiplier for inputs and outputs
ν_g	Weight or multiplier of cost objective g
$\tilde{\nu}_\tau$	Cost multiplier for inputs and outputs
$\mathbb{N}$	Set of natural numbers
ϖ_0	Variable used to indicate returns to scale in BCC model
φ^0_*	Optimal value of CCR-O model for DMU under investigation
ψ^0	Super-effectiveness score for DMU under investigation
$\mathbb{R}$	Set of real numbers
Θ	Goal-oriented efficiency
θ^0	Optimal value of GBCC-MIN model for DMU under investigation
θ^0_*	Optimal value of BCC-I model for DMU under investigation
$up_{i,t}$	Slope of upper boundary of intersection $C_{i,t}$ of multicone and $y_i y_t$-plane
w_i^ρ	Relative variation of output i of DMU ρ
X^ρ	Input vector X of DMU ρ
x_j^ρ	Value of input j of DMU ρ
Y^ρ	Output vector Y of DMU ρ
$\bar{y}_i$	Arithmetic mean of output i
y_i^ρ	Value of output i of DMU ρ

1 Introduction

1.1 Motivation and contribution of the dissertation

Organizations are under continuous pressure to improve their and their subunits' performance in order to operate successfully in today's rapidly changing and highly competitive business environment. Before any organization is able to reveal potential for performance improvement, it needs to measure and evaluate the actual level of its organizational performance.[1] Therefore, the attention of organizations' management started to focus on measuring and monitoring systems that collect, generate, and analyze relevant data on performance. Most of such systems capture solely the financial aspects of performance, since financial data is already available from management accounting systems.[2] In order to draw a more comprehensive picture of the organizational performance, non-financial aspects of performance, such as customer satisfaction and product quality, should be considered as well.[3] The challenge herein lies in a set of performance measures defined in different units of measurement, which make it difficult to assess the overall performance.

Against this background, an analytic technique capable of taking into account numerous performance measures gained considerable interest of academic research in the field of performance measurement and management: data envelopment analysis (DEA). The attractiveness of DEA lies in its ability to evaluate the relative performance of comparable organizations or organizational subunits by aggregating all relevant performance measures – independent of their units of measurement – into a single comprehensive efficiency score. Moreover, for each underperforming – inefficient – unit, DEA suggests an individual path to improve its efficiency through detecting appropriate benchmarks along with possible improvement potential specific to each performance measure.[4]

The mathematical background of DEA methodology offers a considerable flexibility and potential for further methodological development. On the one hand, this has led to an enormous growth of the DEA research, resulting in various

1 Cf. Neely (1999), p. 213.
2 Cf. Bredrup (1995a), p. 77; Bredrup (1995b), p. 169; Ghalayini & Noble (1996), p. 64; Lohman et al. (2004), p. 269; Brignall (2008), p. 161; Nudurupati et al. (2011), p. 282.
3 Cf. Kaplan & Norton (1992), p. 71; Waggoner et al. (1999), p. 54; Bourne et al. (2000), p. 754.
4 Cf. Paradi & Sherman (2014), pp. 33–34.

extensions of the DEA methodology, as demonstrated by more than 4,000 publications on the subject of DEA.[5] On the other hand, the understanding of the mathematical background of the DEA evaluation procedure appears to be very challenging for the non-scientists, such as practitioners.[6] Without the support of DEA experts, e.g. DEA researchers or scientists, practitioners are unable to cope with an appropriate application of DEA for their organizations. This confirms the limited use of DEA in business practice, which in turn is difficult to accept for DEA scientists, like Paradi & Sherman (2014), who stated that

> [t]he limited use of DEA in practice is frustrating to those familiar with its power. This technique has the ability to guide organization to be more efficient, reduce operating costs, and improve profitability in ways that are not possible with other methods.[7]

In order to close the gap between DEA research and its use in business practice, practitioners' requirements should be given proper attention within the DEA research.[8] For this purpose, collaboration between scientists and practitioners is inevitable, as noticed by Liu et al. (2013b),

> [f]urther development of DEA requires cooperation between theoreticians and practitioners. Only with the help of theoreticians can practitioners put the method into practice to meet their needs, and with the feedback from practitioners, theoreticians can refine their theories to make further contributions.[9]

Accordingly, it can be stated that there is a need for methodological development of DEA, which is driven by the demands of modern organizations, i.e. an "applications driven theory"[10] is needed. This forms the motivation of the underlying dissertation.

The "applications driven theory" implies the interaction with practitioners. For this purpose, a research project carried out by the Institute of Management Control and Business Accounting of the Technische Universität Braunschweig – in close collaboration with a pharmacy chain that operates in a European country – was used as a basis. The challenges based on government regulations[11] applied to

---

5 Cf. Emrouznejad et al. (2008), p. 152; Liu et al. (2013a), p. 3.

6 Cf. Ozbek et al. (2009), p. 830.

7 Paradi & Sherman (2014), p. 30.

8 Cf. Cooper (2014), pp. 8–9; Liu et al. (2013b), p. 900; Paradi & Sherman (2014), p. 48.

9 Liu et al. (2013b), p. 900.

10 Cooper (2014), p. 9.

11 The most common types of government regulations currently applied to pharmacy sector in European countries concern: restrictions on pharmacy ownership and the establishment of pharmacies, price and pharmacy remuneration controls, restrictions

the pharmacy sector in most European countries and increased public interest in health care force the management of the pharmacy chain attaching considerable importance to the performance evaluation and control of its pharmacies. The approach for performance evaluation currently used by the management was based on ratio analysis of multiple key performance indicators (KPIs). However, the various ratios of KPIs usually provide different – sometimes even contradictory – statements about pharmacies' performance, impeding assessment of the overall performance of each pharmacy. Therefore, the management of the pharmacy chain was willing to try out the DEA for performance evaluation.

From scientific point of view, the purpose of the project was twofold: firstly, to analyze the difficulties in applying DEA to the particular pharmacy chain. Secondly, to gain insight into the managerial decision making of this organization, in order to detect requirements of business practice that should be taken into account when providing new methodological developments of DEA. Concerning the former aspect, the need for an applicable framework for performance evaluation based on DEA that elucidate DEA's potential for performance management was detected. This finding provides the starting point for the definition of a comprehensive framework for performance evaluation based on DEA, embedding thereby DEA in an integrated concept of performance management. Concerning the second aspect, information acquired within the above mentioned project facilitated the identification of three issues of managerial relevance within the developed framework. These issues promote methodological developments of DEA, which allow a multidimensional point of view on the evaluation of organizational performance.

The major contributions of the dissertation are:

- the definition of a comprehensive and systematic step-by-step framework for performance evaluation based on DEA as one part of a performance management process,
- the extensions of DEA methodology for the evaluation of two additional performance dimensions – effectiveness and balance – motivated by the intention to meet the demands of business practice and to rectify some of the limitations of DEA, and
- the development of a goal-oriented approach to determine suitable performance criteria for DEA efficiency evaluation, referring to the extension of DEA methodology explicated by Dyckhoff & Ahn (2010) as generalized DEA (GDEA).

---

on distribution to pharmacy outlets as well as registration and licensing of pharmacies. For a detail description of the government regulations, see Taylor et al. (2004), Vogler et al. (2006), and Volkerink et al. (2007).

The information and data obtained from the pharmacy chain is the basis for a consistent example used throughout the dissertation. However, for the purpose of confidentiality, all commercial sensitive information – that would enable an identification of the pharmacy chain – is removed and original data is modified. Furthermore, if parts of required information were not available, adequate fictive substitutes were generated in order to enable a presentation of the introduced DEA approaches.

1.2 Structure of the dissertation

An outline of the dissertation's structure is presented in Figure 1.1. The corresponding explanations are provided below.

Chapter 2 provides a conceptual framework for performance evaluation based on DEA, focusing on the clarification and delineation of key terms concerning the management, measurement, and evaluation of performance. The analysis of the framework for possible potential of methodological developments of DEA reveals three issues entailing managerial problems. For each of them a research question is derived and addressed in detail in one of the following chapters. This is done from a multidimensional point of view on the evaluation of organizational performance.

Chapter 3 presents an overview of the conceptual disorder in the pertinent literature concerning the definition and evaluation of effectiveness. Characterizing effectiveness as an independent performance dimension, which refers exclusively to the original goals of economic activities, a new DEA approach for the evaluation of effectiveness is developed. The approach is capable of incorporating information about predetermined goals and corresponding target levels already available in the context of common performance management systems.

Chapter 4 addresses the managerial problems resulting from flexible weights in DEA that allow the exclusion of performance criteria from the evaluation. Therefore, it introduces an innovative DEA approach that considers balance in criteria achievement as an additional dimension of performance. The resulting balance score represents a complementary measure to traditional efficiency and effectiveness scores, providing important insights into the performance structure of the organization or organizational subunit under investigation.

Chapter 5 outlines the problems in the performance criteria determination required for the DEA efficiency evaluation, revealing the need to expand the DEA methodology by GDEA. Particular emphasis is given to the introduction of a stepwise approach that assists decision makers in determining suitable and measurable performance criteria from a goal-oriented perspective. Finally, Chapter 6 summarizes the main findings of the dissertation and proposes directions for further research.

Fig. 1.1: Structure of the dissertation

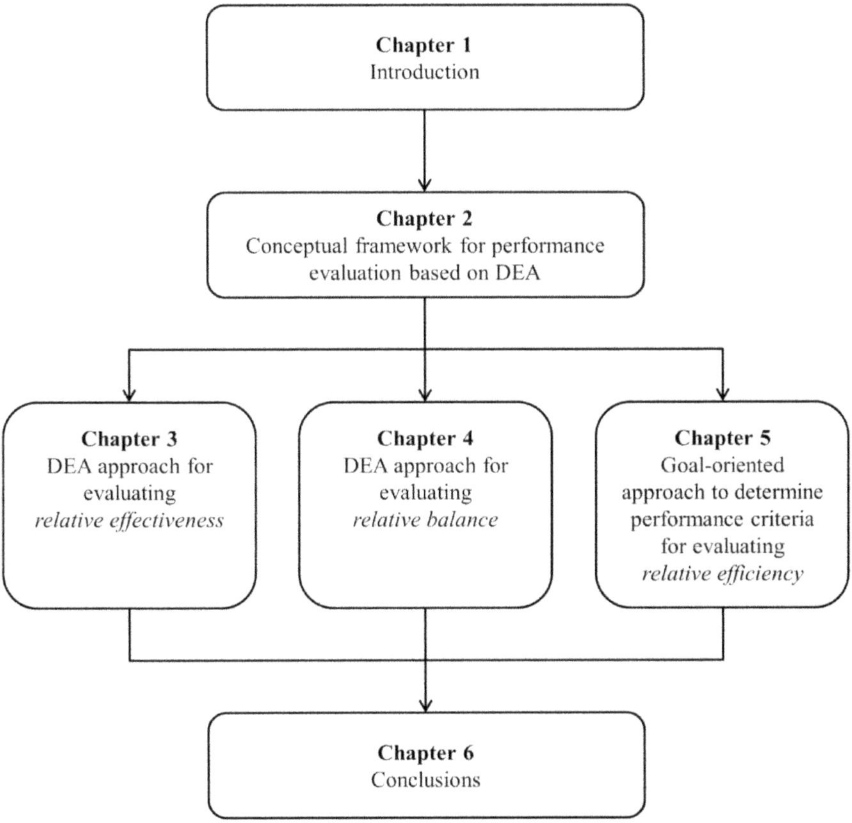

2 Conceptual framework for performance evaluation based on DEA

Organizational performance as well as its measurement and management have already been attracting significant interest both from scientists and practitioners for several decades.[12] However, this rather results in ambitious and eclectic concepts, instead of a commonly accepted general theory on performance management and measurement. Against this background, the aim of this chapter is to provide an integrated concept of performance management that clearly differentiates between performance measurement, performance evaluation, and performance management. Moreover, since performance evaluation is detected as the core element of performance management, particular emphasis is given to the development of an adaptable, comprehensive, and systematic framework for performance evaluation. For this purpose, *data envelopment analysis* (DEA) – an analytic technique capable of covering the complexity of modern organizations and their numerous performance measures, on the one hand, and facilitating managerial decision making by suggesting ways for future performance improvement, on the other – is incorporated into the process of performance evaluation. Finally, in order to advance the use of the proposed framework in practice, three issues of managerial relevance within the developed framework are uncovered, promoting methodological developments of DEA, which allow a multidimensional point of view on the evaluation of organizational performance.

This chapter is organized as follows. Section 2.1 introduces the key concepts and terms. It starts with performance, differentiates between performance measurement and performance evaluation, and presents a comprehensive and systematic process of performance management, in which performance evaluation is shown an essential element. Section 2.2 provides an introduction into the DEA methodology, presenting a brief overview of the basic DEA models, examining the strengths and limitations of DEA from a managerial point of view, and describing a systematic procedure for DEA applications designed by Golany & Roll (1989). In Section 2.3, the comprehensive and systematic step-by-step framework for the performance evaluation based on DEA is developed through the integration of the DEA application procedure of Golany & Roll (1989) into the process of performance evaluation, which is briefly described at the beginning of this section.

12 Cf. Venkatraman & Ramanujam (1986), p. 801; Lebas & Euske (2007), p. 125.

Subsequently, further potential of the proposed framework for methodological developments of DEA is detected, providing the basis for the definition of the research questions of this dissertation. Section 2.4 concludes and discusses the chapter.

2.1 Introduction of key concepts and terms

2.1.1 Performance: an ambiguous term

Performance has become a recurrent theme in the management literature as well as business practice.[13] Both researchers and practitioners are actively working on the meaning and measurement of the performance of organizations in order to reveal potential for performance improvement. Although the volume of the literature on these topics continually increases, a clear definition of performance can hardly be found.[14] Lebas & Euske (2007), e.g., stated that

> [d]espite the frequency of use of the word [performance], its precise meaning is rarely explicitly defined by authors, even when the main focus of the article or book is performance.[15]

A good explanation for that was given by Folan et al. (2007),

> [p]erformance – we have come to think, often unconsciously – needs no introduction as everyone knows what performance is.[16]

The lack of an adequate definition of performance along with its widely use in different fields of management results in a diversity of meanings of performance. For instance, it is often considered as interchangeable with terms such as productivity, profitability as wells as efficiency and effectiveness.[17] Accordingly, it cannot be simply assumed that the meaning of performance is obvious.[18] Instead, it seems that performance is an "ambiguous term"[19] allowing various interpretations,[20] since – as pointed out by Lebas & Euske (2007) – performance is

---

13 Cf. Folan & Browne (2005), p. 663; Neely (2007a), pp. 1–3.
14 Cf. Neely et al. (1995), p. 80; Richard et al. (2009), p. 719.
15 Lebas & Euske (2007), p. 125.
16 Folan et al. (2007), p. 605.
17 Cf. Bredrup (1995b), p. 173; Neely et al. (1995), p. 80; Tangen (2005a), p. 34; van Helden & Reichard (2013), p. 11.
18 Cf. Folan et al. (2007), p. 605.
19 Otley (1999), p. 364.
20 Cf. Otley (2001), p. 251; Folan et al. (2007), p. 609; Lebas & Euske (2007), p. 131; Bovaird (2008), p. 185.

> [...] one of those "suitcase words" [...] in which everyone places the concepts that suit them, letting the context take care of the definition.[21]

Despite the missing agreement on a common definition of performance, there is a consensus about the main characteristics of performance, as illustrated in Figure 2.1. In the following, these characteristics are briefly presented.

- Performance does not have a unique definition. It is *contextual* concerning the situation and the intention of the persons, such as decision makers, who want to evaluate performance of an organization, a company, a process, a product or an individual.[22] Thereby, performance is usually related to the relevant properties, activities, and/or objectives of the corresponding organization or organizational subunit under consideration.[23]
- Performance is *multidimensional.*[24] Performance encompasses all concepts that detect the economic and operational benefit and success achieved by the respective organization or organizational subunit under consideration. Performance dimensions may vary between productivity in a manufacturing process, monetary profitability, economic use of resources – when providing (producing) services (goods) – and/or the extent of goal achievement.[25]
- Performance is a *relative* concept. As mentioned above, performance is not of an absolute nature and does not have a standardized definition. It is rather closely associated with the meaning of progress or the potential of value creation. Hence, a comparison relative to a reference, e.g. a predetermined target, is required to quantify the performance.[26] The choice of a reference as well as the definition of relevant performance criteria is decisive for the evaluation of performance.
- Performance is *measurable* by means of relevant performance criteria, which are selected to serve as surrogates.[27] According to the specific context, these criteria may include financial as well as non-financial measures, e.g. customer satisfaction. The only condition is that they should best capture the performance of the organization or organizational subunit under consideration.

---

21 Lebas & Euske (2007), p. 126.
22 Cf. Lebas (1995), pp. 24–29; Tangen (2005a), p. 40; Lebas & Euske (2007), p. 133.
23 Cf. Meyer (2002), p. 9.
24 Cf. Bhargava et al. (1994), p. 245; Neely (1999), p. 222; Tangen (2005b), p. 48; Brignall (2008), p. 160; Richard et al. (2009), pp. 722–723; Bentes et al. (2012), p. 1791.
25 Cf. Neely et al. (1995), p. 80; Tangen (2005a), pp. 35–40.
26 Cf. Folan et al. (2007), pp. 608–610; Lebas & Euske (2007), p. 136.
27 Cf. Lebas (1995), pp. 25–26; Wholey (1996), p. 147; Folan et al. (2007), p. 612; Lebas & Euske (2007), p. 134.

- Performance is *subjective*. Each person, such as a decision maker or a stakeholder, is able to influence the definition and interpretation of performance, e.g. through the selection of different relevant performance criteria from her or his own perspective.[28]

Fig. 2.1: Main characteristics of performance

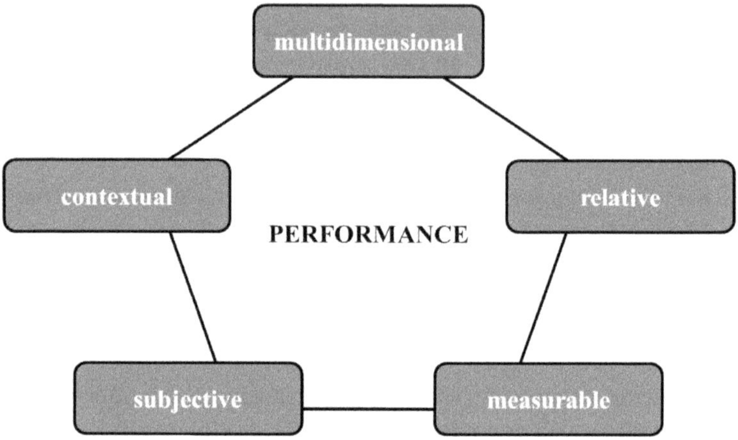

In consideration of the above characteristics of performance, it can be concluded that performance is a complex concept, which is case and user specific. The users, e.g. decision makers, decide about

- the context in which the performance is measured,
- the dimensions of performance that are involved,
- the reference relative to which performance is compared,
- the relevant criteria that depict the performance appropriately, providing the basis for the quantification of performance and
- the subjective judgments that are incorporated.

Consequently, the concept of performance has to be operationalized in each specific case – concerning at least the above mentioned characteristics – before it can be measured and managed.[29]

---

28 Cf. Lebas (1995), p. 25; Wholey (1996), p. 146; Folan & Browne (2005), p. 664; Folan et al. (2007), pp. 612–614; Lebas & Euske (2007), pp. 131–132.

29 Cf. Lebas (1995), p. 24.

2.1.2 From performance measurement to performance evaluation

Measurement serves as a starting point for the improvement of performance.[30] It quantifies the actual level of organizational performance and thereby constitutes a source of useful information that supports managerial decision making.[31] As Sir William Thomson, 1st Baron Kelvin (1889) rightly stated,

> [...] when you can measure what you are speaking about, and express it in numbers, you know something about it [...] [otherwise] your knowledge is of a meagre and unsatisfactory kind: it may be the beginning of knowledge, but you have scarcely, in thoughts, advanced to the stage of science, whatever the matter may be.[32]

It is thus hardly surprising that most organizations have a *performance measurement system*[33] (PMS) of some kind.[34] Such systems are primarily dominated by financial data available from management accounting systems, which constitute a basis for economic judgments and decisions.[35] However, the measurement of solely financial aspects of performance is insufficient in today's rapidly changing and highly competitive business environment. It provides an incomplete picture of organizational performance.[36] For long-term survival and business success of an organization, non-financial aspects of its performance, e.g. product quality, innovation, customer and employee satisfaction, etc. – measured outside the traditional management accounting systems – have to be considered as well.[37]

The dissatisfaction with accounting based measurement systems has been reflected by the extensive discussion and documentation of their shortcomings, such

---

30 Cf. Lohman et al. (2004), p. 267; Spitzer (2007), p. 1; Barnes & Radnor (2008), p. 103; Brignall (2008), p. 160; Grafton et al. (2010), pp. 692–694; Taticchi et al. (2010), p. 4.

31 Cf. Lohman et al. (2004), p. 271.

32 Thomson (1889), pp. 73–74.

33 Tangen (2005b), p. 46, defines performance measurement systems as „a set of performance measures (i.e. metrics used to quantify the efficiency and effectiveness of action) that provides a company with useful information that helps to manage, control, plan and perform the activities undertaken in the company".

34 Cf. Bredrup (1995b), p. 169.

35 Cf. Bredrup (1995a), p. 77; Bredrup (1995b), p. 169; Ghalayini & Noble (1996), p. 64; Lohman et al. (2004), p. 269; Sprinkle & Williamson (2007), p. 428; Brignall (2008), p. 161.

36 Cf. Pekuri et al. (2011), p. 50.

37 Cf. Kaplan & Norton (1992), p. 71; Bredrup (1995a), pp. 78–85; Ghalayini & Noble (1996), p. 64; Waggoner et al. (1999), pp. 53–54; Bourne et al. (2000), p. 754; Ittner & Larcker (2003), p. 88; Nudurupati et al. (2011), pp. 280–282; Kasie & Belay (2013), pp. 599–600.

as short termism and internal focus of the measures, inability to forecast the future, lack in strategic focus, and lack in continuous improvement.[38] These criticisms have encouraged the search for new measurement systems and have thereby triggered a "revolution"[39] in the development of performance measurement. Numerous researchers from a wide range of disciplines – starting from operations and production management through accounting and management control, information systems and operations research to human resource management and psychology – are exploring the subject.[40] The fascinating abundance[41] of the resulting publications on performance measurement creates both opportunities as well as challenges.[42] On the one hand, various new PMSs emerge to facilitate the alignment between business strategy and performance measures with aiming to capture almost all aspects of organizational performance.[43] On the other hand, performance measurement remains a broad topic without a cohesive body of knowledge because the heterogeneity of application and research areas hinders the development of a common theoretical foundation.[44] Rouse & Putterill (2003) claimed that after more than 40 years of research, an elaboration of a single common framework for performance measurement is impossible. Moreover, Franco-Santos et al. (2007) reported that there is only little agreement on the main characteristics and purposes of PMSs, varying thereby from the measurement and information provision to evaluation, progress monitoring, and internal communication. Such diversity of purposes or functions of performance measurement creates confusion not only in theory but also in practice.[45] Therefore, it is necessary to clarify the meaning of performance measurement and its role in the management of performance.

38 A summary of the criticisms concerning the traditional measurement systems based on financial data is provided in Ghalayini & Noble (1996), pp. 65–67; Bourne et al. (2000), pp. 754–755; Tangen (2004), pp. 726–727; Melnyk et al. (2014), p. 176.

39 Neely (1999) composed an article on the topic: "The performance measurement revolution: why now and what next?"

40 Cf. Marr & Schiuma (2003), p. 684; Neely (2005), p. 1268; Neely (2007a), p. 2; Taticchi et al. (2010), p. 5.

41 The citation analysis carried out by Taticchi et al. (2010) reveals 6,618 papers on performance measurement in the ISI Web of Knowledge database for the time period 1970–2008.

42 Cf. Neely (2007b), p. 7.

43 A chronological review of PMSs along with their key characteristics is provided in Pun & White (2005), Taticchi & Balachandran (2008), and Neely et al. (2007).

44 Cf. Neely et al. (1995), pp. 80–81; Marr & Schiuma (2003), p. 684; Neely (2005), pp. 1268–1269; Micheli & Manzoni (2010), p. 466.

45 Cf. Franco-Santos et al. (2007), p. 799; Pavlov & Bourne (2011), p. 104.

One of the reasons for the wide range of purposes associated with performance measurement originates from the recognition that the sole quantification of performance is insufficient for managing an organization,[46] as stated by Amaratunga & Baldry (2002),

> [r]esults of performance measurement indicated what happened, not why it happened, or what to do about it.[47]

More recently, Neely (2007c) clearly summed up that

> [...] measurement itself delivers little value until the data generated are analyzed and acted upon.[48]

Therefore, the primary concern of research shifts from simple measurement towards a detailed analysis and interpretation of the results in order to facilitate managerial decision making.[49] This means that aspects such as assessment and evaluation of performance become the focus of increased attention. However, contrary to the widespread view, these aspects are here not seen as parts of performance measurement. Instead, performance measurement, performance evaluation as well as performance assessment are characterized as dependent but different activities.[50] Performance measurement is narrowed to the non-judgmental activity of quantifying the current level of performance.[51] The results delivered by the underlying PMS provide the basis for the assessment and evaluation of performance.[52] Obviously, performance measurement is an essential prerequisite of performance assessment and performance evaluation.[53]

The terms performance assessment and performance evaluation are often used equivalent to each other.[54] However, according to Afsharian et al. (2014), there is a clear distinction between them. Performance assessment focuses on the interpretation of the results provided by performance measurement,[55] e.g. the assessment of the gap between current and desired performance, the identification of areas of

46 Cf. Lebas (1995), p. 35; Melnyk et al. (2014), p. 175.
47 Amaratunga & Baldry (2002), p. 218.
48 Neely (2007c), p. 75.
49 Cf. Folan et al. (2007), p. 610; Bititci et al. (2012), p. 314.
50 Cf. Bredrup (1995b), p. 175; Spitzer (2007), p. 41.
51 Cf. Bredrup (1995b), pp. 169–174; Spitzer (2007), p. 42.
52 Cf. Bredrup (1995c), p. 191; Amaratunga & Baldry (2002), p. 218; Spitzer (2007), pp. 41–42.
53 Cf. Rouse & Putterill (2003), p. 796.
54 Cf. Folan et al. (2007), p. 608.
55 Cf. Afsharian et al. (2014), p. 254.

strengths and weaknesses, the provision of the basis for decisions concerning the distribution of incentives and/or penalties as well as corrective actions. In contrast, performance evaluation is understood as a broader concept in terms of a judgmental process, which encompasses performance measurement and performance assessment,[56] as illustrated in Figure 2.2. The aim of performance evaluation consists in supporting an organization in monitoring its performance in relation to its goals.[57] To ensure a proper evaluation of organizational performance, it is, furthermore, of critical importance to prepare a framework prior to the measurement and assessment, i.e. to determine relevant performance criteria and a PMS by considering the context and goals for which the performance is evaluated.[58]

Fig. 2.2: The concept of performance evaluation[59]

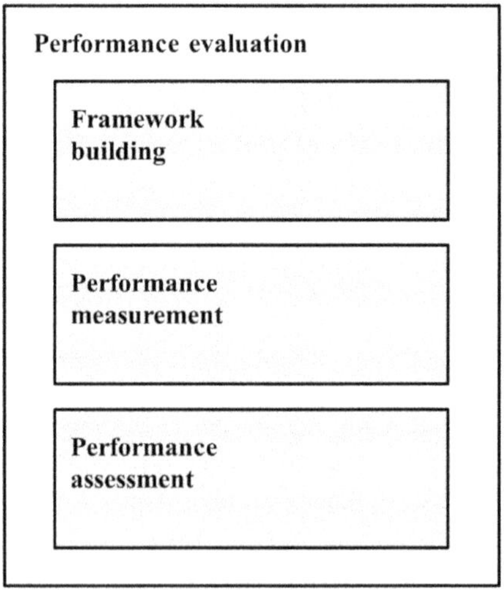

Finally, performance evaluation not only incorporates managerial information by establishing a link between the goals of an organization and the measurement of its performance, but also assesses and feeds back relevant information to facilitate

56 Cf. Rouse & Putterill (2003), pp. 801–802; Afsharian et al. (2014), p. 254.
57 Cf. Bruns & McKinnon (1992), p. 17; Chen et al. (2011), p. 909.
58 Cf. Afsharian et al. (2014), p. 254.
59 Cf. Afsharian et al. (2014), p. 255. For a detailed description of the concept of performance evaluation according to Afsharian et al. (2014), see Subsection 2.3.1.

managerial decision making and control.[60] Accordingly, it represents an essential element in the management process of organizational performance.[61]

2.1.3 Performance management: an umbrella concept

The concept of performance management – like performance measurement – has caused an explosion of interest from a wide range of disciplines, resulting in a diversity of views on what constitutes performance management.[62] In recent years, the concept of performance management becomes closer to strategic management of organizations.[63] Due to rapidly changing and consumer-led markets, organizations attach greater importance to the implementation of a suitable business strategy, which is continuously adjusted and reformulated in order to ensure the long-term business success of the corresponding organization. Such a strategy is usually reflected by a set of harmonized purposes and the respective goals. In this context, performance management supports the alignment of organizational performance with the strategic goals of an organization.[64] Therefore, it encompasses all managerial activities that are necessary to ensure the congruence between the performance of an organization – and of its employees – and the organizational strategic goals.[65] Otley (2001) argued that

> [...] performance management provides an umbrella under which we can study the more formal processes that organizations use in attempting to implement their strategic intent, and to adapt to the circumstances in which they have to operate.[66]

The characterization of performance management as an umbrella concept indicates a broad understanding of the topic. Thus, it is not surprising that the elements of performance management mentioned in the literature are as eclectic as the diversity of views on performance management.

Although a holistic definition of *performance management process* (PMP) does not currently exist, the existing concepts demonstrate a similar understanding of its key elements: planning, measurement, assessment, and review. Differences merely exist in the degree of detail of particular elements of PMP. For instance,

60 Cf. Sleezer et al. (1998), pp. 63–66.
61 Cf. Marques et al. (2011), p. 1058.
62 Cf. Smith & Goddard (2008), p. 121.
63 Cf. Bititci et al. (1997), p. 46; Radnor & Barnes (2007), p. 392; Smith & Goddard (2008), p. 120; Adler (2011), p. 255.
64 Cf. Aguinis (2013), pp. 2–3.
65 Cf. Bititci et al. (1997), p. 47; Aguinis (2009), p. 5.
66 Otley (2001), p. 250.

Bredrup (1995a) suggested three main elements of PMP – planning, improvement, and review – whereas Aguinis (2009, 2013) distinguished between six elements of PMP – prerequisites, planning, execution, assessment, review, and renewal and re-contracting. The various concepts of PMP have one thing in common: their elements can be assigned to the typical functions of management – planning and control.[67] Additionally, Ahn (2003) suggested that learning is an additional function of management and should, therefore, be part of a typical management process. Accordingly, the underlying dissertation adopts this view, but additionally suggests inserting performance evaluation as an intermediate element of PMP between performance planning and performance control. The resulting PMP is illustrated in Figure 2.3.

Fig. 2.3: Performance management process

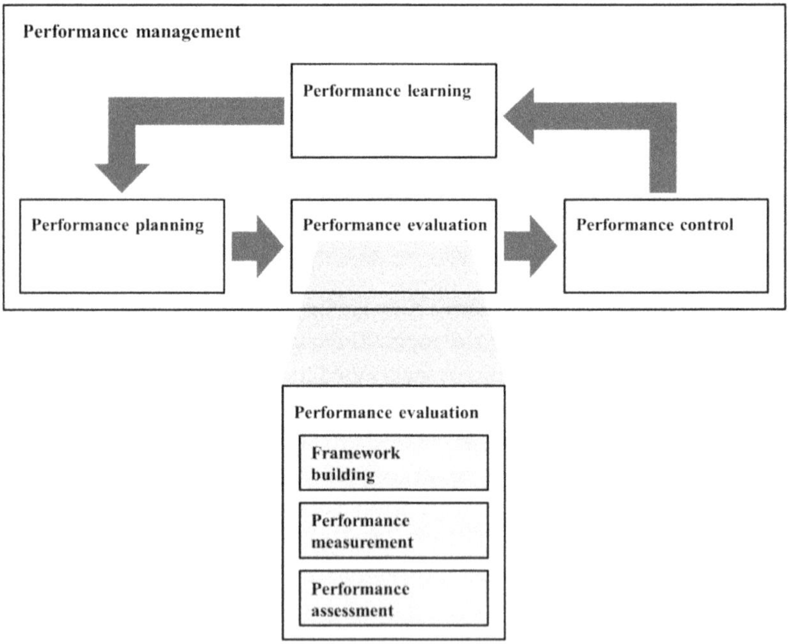

Performance management starts with performance planning. The aim of performance planning is to link the strategy of an organization with the underlying concept of performance. Therefore, it is necessary to structure and clarify the

67 Cf. van Helden & Reichard (2013), p. 11.

organizational strategic goals with regard to the relevant dimensions of performance. Once the goals and the corresponding targets have been determined, they have to be communicated through the organization to ensure that not only the managers but also the employees are aware of these goals and, consequently, align their performance on the achievement of these goals. Furthermore, on the management level decisions have to be made concerning the allocation of resources required and actions to be taken in order to accomplish the organizational goals.

To detect whether the organization is successful in aligning its performance in accordance with its goals, it is necessary to measure and assess its actual performance. This is the focus of the next step of PMP – the performance evaluation. As an intermediate element of PMP, it incorporates all relevant information from the previous step of PMP – the performance planning. Thereby, it determines measurable performance criteria for each goal and selects a suitable PMS. Based on this preliminary framework, the current level of organizational performance can be quantified and assessed whether there are significant performance gaps and/or areas for improvement of performance. The information gained within this step is crucial for the subsequent managerial decision making. Therefore, it is fed forward to the next step of PMP – the performance control.

The aim of performance control is to look backwards and analyze the alignment of organizational performance with regard to its goals. Besides the distribution of incentives and/or penalties according to the results of performance evaluation, it focuses especially on the examination of the reasons for performance gaps – if detected – and the implementation of appropriate corrective actions to eliminate them in the short term. In order to ensure the achievement of organizational goals and sustainability of high performance in the long term, all insights gained within this step are fed forward to performance learning, as the next and final step of PMP.

The emphasis of performance learning lies in the use of internal information, i.e. the findings obtained from the previous steps of PMP, and external information, e.g. the opinions and experiences of third parties about the performance of the organization, to learn from the past decisions and experiences for the future.[68] As a consequence, the knowledge gained within this step affects and influence future decision making, e.g. concerning the adjustment of planned goals and/or chosen performance criteria – if necessary. Therefore, it is fed forward to performance planning, implying that the PMP is a cycle that starts all over again.[69]

---

68 Cf. Ahn (2003), p. 84.

69 Cf. van Helden et al. (2012), p. 162; Aguinis (2009), p. 25; Aguinis (2013), p. 53.

Consequently, performance management is as a continuous[70] process intended to support an organization in monitoring its performance in relation to its strategic goals,[71] by simultaneously increasing the rationality of managerial decision making.[72] On the one hand, the comprehensive systematic process of performance management links strategic goals to pertinent routine performance information. In doing so, it allows the evaluation and analysis of the goal attainment in light of current organizational performance.[73] On the other hand, the PMP entails both a feedback loop that reports the crucial results of performance evaluation to the decision makers with regard to the control of the current (past) performance, and a feed forward loop that stimulates organizational learning from the impact of past decisions, by facilitating goal setting and development of action plans in order to enhance the organization's future performance.[74]

Obviously, the closed loop concept of performance management is reliant on the gathering, analysis, and provision of all relevant performance information.[75] Performance evaluation focuses exactly on these issues, representing thereby the core element of the PMP. Therefore, the emphasis of the underlying dissertation is the development of an adaptable, comprehensive, and systematic performance evaluation framework that fulfills its intermediate role as information supplier, capturing the multiple performance dimensions and performance criteria in accordance with the goals of an organization. With respect to this purpose, the next section introduces the data envelopment analysis – an analytic technique capable of covering the complexity of modern organizations with respect to their numerous performance measures.

2.2 Data envelopment analysis (DEA)

2.2.1 Introduction of DEA

DEA is recognized as a powerful tool in the relative evaluation of efficiency, as the most widely used performance dimension, with a broad spectrum of applications. The value of DEA lies in its capability to evaluate the relative efficiency[76]

70 Cf. Aguinis (2009), p. 5; Aguinis (2013), p. 2.
71 Cf. Bititci et al. (1997), p. 47.
72 Cf. van Helden et al. (2012), p. 161.
73 Cf. Amaratunga & Baldry (2002), pp. 220–222; van Helden et al. (2012), p. 161.
74 Cf. Grafton et al. (2010), pp. 692–693.
75 Cf. Bititci et al. (1997), p. 47.
76 In this context, efficiency is a synonym of productivity that is expressed as a ratio of produced outputs to consumed inputs (cf. Farrell, 1957). Farrell's (1957) seminal con-

of comparable decision making units (DMUs) by aggregating all relevant inputs, e.g. resources used, and outputs, e.g. goods or services produced, – i.e. performance criteria – into a single comprehensive efficiency score.[77] Typically, DMUs represent public institutions, e.g. hospitals or universities, or profit organizations of the same sector, e.g. energy distribution companies, as well as branch stores or divisions of the same company, e.g. bank branches. To be comparable, the DMUs have to fulfill the assumption of homogeneity, i.e. they have to operate in similar environments by using similar resources or activities – as inputs – to produce comparable goods or services – as outputs.[78] The measures of inputs and outputs do not need to be in the same metric, i.e. to have the same units of measurements.[79] This allows the use of both financial and non-financial measures for the evaluation of efficiency.

As a non-parametric approach, DEA does not require any specifications of both a particular production function and weights for the inputs and outputs in advance of the efficiency evaluation. Instead, the actual data set provides an empirically constructed production/efficient frontier, which is formed by best performing, i.e. efficient, DMUs in the sample. The different linear facets of this frontier connect the efficient DMUs with each other and envelop the remaining, i.e. inefficient, DMUs. Consequently, the inefficient DMUs are evaluated relative to the best practices located on the frontier. The radial projection of an inefficient DMU onto the efficient frontier needs the identification of a reference point as a benchmark against which such inefficient DMU can be compared. This reference point could be one of the efficient DMUs or a convex combination of several efficient DMUs, defining a virtual DMU.[80] The distance of inefficient DMU to its benchmark represents the level of its inefficiency. Figure 2.4 illustrates the idea in the case of one input and two outputs. Here, the efficiency frontier, formed by the efficient DMUs B, C, and D, envelops the inefficient DMUs A and E. The broken lines through these DMUs onto the frontier illustrate the radial projection required to identify the appropriate benchmarks, A* – as a convex combination of DMUs B and C – and E* – as a convex combination of DMUs C and D.

---

tribution provided the starting point for efficiency and productivity analysis and thus for the introduction of DEA by Charnes et al. in 1978 (cf. Paradi & Sherman, 2014).

77 According to Ahn & Le (2014), p. 4, there is a distinction between input and output factors – "as objects from a stock-oriented view" – and inputs and outputs, which in a dynamic context are used to classify factors that enter a production process or result from a production process. In the following, it is not distinguished between the terms "input" and "input factor" as well as between "output" and "output factor".

78 Cf. Golany & Roll (1989), p. 239; Dyson et al. (2001), p. 247; Thanassoulis (2001), p. 21.

79 Cf. Cooper et al. (2007b), p. 22.

80 Cf. Cooper et al. (2007b). pp. 7–9.

Fig. 2.4: Efficiency frontier in the case of one input and two outputs

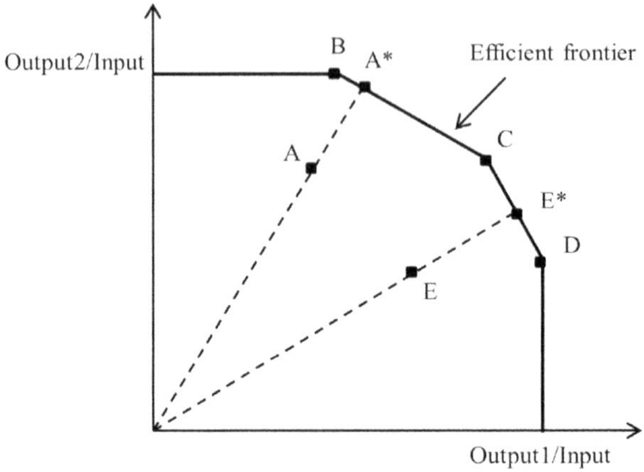

The weights for the performance criteria are also dependent on the underlying empirical data set. They are endogenously determined within the evaluation by choosing those weights for each DMU, which show this DMU in the best possible light relative to the other DMUs, i.e. given the same weights no other DMU would be more than 100% efficient. The calculated pareto optimal efficiency scores[81] are scaled to a maximum value of 1 (or 100%) for efficient DMUs. Accordingly, inefficient DMUs obtain scores between 0 and 1 (or 0% and 100%). The difference between the current (in)efficiency score of a DMU and full efficiency of 1 (or 100%) indicates by which amount this DMU should decrease its inputs (in case of input orientation) or increase its outputs (in case of output orientation) in order to become at least weak efficient.[82] Weak efficiency means that although the corresponding DMU is by now part of the efficient frontier, there is still additional – non-radial – improvement potential concerning the input excesses and/or output shortfalls – identified as input/output slacks. Consequently, the DMUs with an efficiency score of 1 (or 100%) and all input/ouput slacks of zero are classified as strong or fully efficient.[83] Therefore, only the strong, i.e. fully, efficient DMUs are referred to as benchmarks for the inefficient as well as for weak efficient DMUs.

81 According to Sengupta (2012), p. 27, a DMU is pareto efficient in producing its outputs from given inputs, if no other DMU or a combination of DMUs can produce more of some outputs without producing less of any other output and without utilizing more of any input.

82 Cf. Yang et al. (2009), p. 207.

83 Cf. Cooper et al. (2007b), pp. 44–45.

Finally, the basic assumptions of the DEA briefly described above represent the conventional DEA methodology, originating in the first decade since the development of the first DEA model in 1978. However, as a mathematical programming technique, DEA offers a considerable flexibility and potential for further methodological developments. This has led to a strong growth of the DEA research, resulting in various extensions of the DEA methodology. In the next subsection, a brief overview of the basic DEA models according to their main different features and characteristics is presented.

2.2.2 DEA models

Since the introduction of DEA in 1978 by the seminal paper of Charnes et al., there has been an enormous growth in methodological developments. Cook & Seiford (2009) provided a comprehensive review of the major theoretical developments during the past three decades. However, the theoretical contributions of various DEA models do not necessarily enhance their application to practical situations. The growing number of DEA models makes it rather difficult to decide which model is appropriate for the particular case. To facilitate the process of selecting a suitable DEA model, Ahn et al. (2014) presented a concept of six general features that enable the fundamental differentiation between DEA models. These features differ in terms of assumptions about the characteristics of DMUs under consideration and the underlying production possibility set. An outline of these features is presented below.[84]

- *Model orientation*
  A distinction is drawn between input-oriented and output-oriented approaches. While input orientation is aimed at producing the same amount of outputs by consuming fewer inputs (minimization of input resources), the output orientation strives for the production of larger quantity of outputs by consuming the same amount of inputs (maximization of output production). Unoriented models do not place such focus on solely minimization of inputs or maximization of outputs.
  The decision regarding the model orientation should take into account to what extent the inputs and outputs are under the direct control of the DMUs. For instance, if DMUs have primary influence over their inputs and less over their outputs, the choice of an input-oriented model is obvious.

84 The last two features – "relevant time period" and "efficiency change over time" – proposed by Ahn et al. (2014) were combined into one due to the mutual dependency between them, reducing the concept to five features.

- *Returns to scale*
  The concept of returns to scale focuses on the relationship between inputs and outputs within the production process. If the outputs increase/decrease by the same proportion as the inputs increase/decrease, the corresponding production function features constant returns to scale (CRS). Otherwise, it refers to variable returns to scale (VRS).
  A production function with the property of VRS can consist of parts or facets where increasing returns to scale (IRS) prevail, i.e. a proportional increase in all inputs results in a more than proportional increase in the outputs, and other parts or facets where decreasing returns to scale (DRS) prevail, i.e. a proportional increase in all inputs results in a less than proportional increase in the outputs. It is also possible that a part of the VRS production function contains CRS (see Figure 2.5).
  The difference between efficiencies calculated on the assumption of VRS and CRS consists in the consideration or rather disregard of scale (in)efficiencies caused by differences in scale size. Since DEA models assigning VRS compare only those DMUs with each other that have similar scale size, the corresponding efficiency scores are already adjusted of scale inefficiencies. Therefore, the number of efficient DMUs under the assumption of VRS is always higher than in case of CRS.[85]

Fig. 2.5: Returns to scale[86]

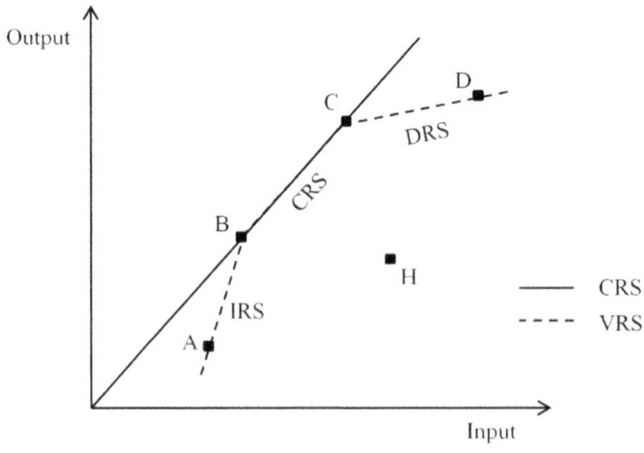

85 For a deeper discussion of returns to scale in DEA, see Banker et al. (2004).
86 Banker et al. (2004), p. 46.

- *Weight restrictions*
  One of the major features of the basic DEA models consists in the complete flexibility of weight determination. Besides the non-negativity restriction on performance criteria, no other specification of weights is required. In some situations, however, the incorporation of additional information, e.g. managerial preferences about the relative importance of particular performance criteria or expert opinion, is crucial for unadulterated performance evaluation. Assignment of absolute or relative restrictions on the weights provides the possibility to take such aspects into consideration.[87]

- *Undesirable performance criteria*
  The production theory usually supposes that outputs as results of the production process are to be maximized, whereas inputs are to be minimized, because their consumption is expensive and therefore undesirable. This assumption is, however, not valid for all kinds of real-world applications. For instance, waste is characterized as an input of waste-burning power plants, but its incineration is welcomed and therefore to be maximized. On the contrary, the resulting CO_2-emission, although an output, is undesirable and accordingly to be minimized. In performance evaluation, such criteria are treated differently as so-called undesirable performance criteria.[88]

- *Relevant time period and efficiency change over time*
  Depending on the available data and the decision about the time focus of the performance evaluation, it is possible to conduct a static (single period) or dynamic DEA (time dependent) evaluation. While static DEA analyses neglect inter-temporal effects in the data, dynamic DEA provides the possibility to discover changes in efficiency over time.
  Moreover, consideration of multiple time periods (panel data) increases the probability that technological changes occur and have an impact on the efficiency of a DMU. In this respect, there are several dynamic DEA approaches that enable the quantification of the technological changes by means of a value.[89]

---

87 An overview of the methods for incorporating weight restrictions in DEA is presented by Thanassoulis et al. (2008).

88 An approach to cope with undesirable performance criteria is proposed by Dyckhoff & Ahn (2010).

89 A brief introduction of most important DEA approaches that deal with time series data is presented by Cooper et al. (2007b), Chapter 11.

In accordance with the above mentioned features, some of the most widely-used DEA models are exemplary classified in Figure 2.6. The presented overview contains 21 alternatives of the following basic DEA models:[90]

- the Charnes-Cooper-Rhodes (CCR) model, Banker-Charnes-Cooper (BCC) model, and additive (ADD) model as the traditional DEA models;
- the assurance region (AR) model and cone ratio (CR) model as prevalent approaches for the incorporation of additional weight restrictions;
- the generalized (GDEA) model for an adequate treatment of undesirable performance criteria;
- the window analysis (Window) and Malmquist productivity index (MPI) for evaluating efficiency over multiple time periods.

Fig. 2.6: Overview of the basic DEA models[91]

		Weight restrictions					
		yes		no			
Returns to scale	constant	AR-I-C CR-I	AR-O-C CR-O	CCR-I	CCR-O	GDEA-C	single (**Time period**)
	variable	AR-I-V	AR-O-V	BCC-I	BCC-O	GDEA-V ADD-V	single (**Time period**)
	constant	–	–	Window-I-C MPI-I-C	Window-O-C MPI-O-C	–	multiple (**Time period**)
	variable	–	–	Window-I-V MPI-I-V	Window-O-V MPI-O-V	–	multiple (**Time period**)
		input	output	input	output	unoriented	
		Model orientation					

90 For a minimal repetition of published material, the following models are only mentioned by name. For full descriptions of these models, see Cooper et al. (2007b). For the description of GDEA, see Dyckhoff & Ahn (2010).

91 Ahn et al. (2014), p. 113. The GDEA models are neither input-oriented nor output-oriented, instead they focus on the minimization or maximization of performance objectives, see Subsection 5.1.3. Therefore, in Figure 2.6, they are referred to as unoriented models.

Each alternative is described by an abbreviation that was previously introduced. Some alternatives have additional abbreviations attached, e.g. (I) for input-oriented and (O) for output-oriented, (C) for constant and (V) for variable returns to scale. Other specifications can easily be read from the illustration. For clarity and due to the fact that only GDEA is able to cope with undesirable performance criteria within the considered selection of basic DEA models, this feature was not explicitly illustrated in Figure 2.6.

2.2.3 Strengths and limitations of DEA

The value-added potential of DEA for performance evaluation and managerial decision making is already well recognized within the international research and scientific community,[92] as demonstrated by more than 4,000 publications on the subject of DEA.[93] However, this may not necessarily be the case for the practitioners, e.g. managers. One of the main reasons for such a disagreement seems to be that managers or rather practitioners have different requirements on model design than the researchers or rather scientists. This affects the use of scientific models in practice.[94] For instance, practitioners usually do not have the time and background to get deep into the model, they prefer to apply models that are much simpler.[95] To encourage the managerial use of scientific models, Little (2004, originally appeared in 1970) suggested to design models that are:[96]

- *Simple*
  The model should be easy to understand.
- *Robust*
  It should be hard to get absurd answers from the model.
- *Easy to control*
  The user should be able to control and influence the model, i.e. to know what input data is required and how to set it to get desired output answers.
- *Adaptive*
  It should be possible to adjust the model as new information becomes available.
- *Complete on important issues*
  All important issues should be included in the model even if they may influence the simplicity and/or require the incorporation of subjective judgments.

---

92 Cf. Jain et al. (2011), p. 617.
93 Cf. Emrouznejad et al. (2008), p. 152; Liu et al. (2013a), p. 3.
94 Cf. Paradi & Sherman (2014), pp. 29–30.
95 Cf. Little (2004), pp. 1842–1843.
96 Cf. Little (2004), pp. 1843–1844.

- *Easy to communicate with*
  The user should be able to rapidly and easily change the input data and obtain and understand the output answers.

In the following, the strengths and limitations of DEA will be critically examined from a managerial point of view, referring to the requirements mentioned above.

The key strength of DEA is its ability to simplify the managerial decision making by solving the problem of multiple and partially contradictory key performance indicators (KPIs), which are usually used in practice, but complicate the assessment of comparable DMUs. DEA aggregates all relevant KPIs into a single comprehensive relative efficiency score, without the need of any a priori specification of both a particular production function and weights for the multiple KPIs or rather performance criteria.[97] However, in order to correctly apply DEA, identify and react to its shortcomings, and analyze the results, it is essential to properly understand the details of the DEA methodology. For the non-scientists, such as practitioners, the understanding of the mathematical background of the DEA evaluation procedure appears to be very challenging, requiring even assistance from DEA experts, such as DEA researchers or scientists. This aspect negatively affects the simplicity of the handling. An adaptable, comprehensible, and systematic step-by-step procedure is needed to guide an inexperienced DEA user through the implementation process of DEA valid for every application case.[98]

As mentioned in Subsection 2.2.1, DEA provides a replicable measure of relative efficiency for each DMU generated from the actual observed data of this DMU in comparison to the best observed value of similar DMUs.[99] Furthermore, DEA offers the possibility for the evaluation of efficiency on a continuous rolling basis, as new data is collected.[100] However, the robustness of the DEA methodology may be compromised, due to its sensitivity to outliers (DMUs with extreme data), measurement errors, missing data, and exclusion of performance criteria from the evaluation.[101] Therefore, it is necessary to carry out

97 Cf. Epstein & Henderson (1989), p. 112; Charnes et al. (1994), p. 8; Ozbek et al. (2009), p. 829.

98 Cf. Paradi & Sherman (2014), pp. 36–37.

99 Cf. Charnes et al. (1994), p. 6.

100 Cf. Jain et al. (2011), p. 617.

101 Cf. Epstein & Henderson (1989), pp. 111–114; Ozbek et al. (2009), p. 830.

a sensitivity analysis of the data used for the efficiency evaluation with respect to these pitfalls.[102]

The focus of DEA on the optimal evaluation of efficiency of each individual DMU rather than of the sample average allows, on the one side, useful insights into the individual performance process of each DMU,[103] e.g. concerning the impacts of every performance criteria incorporated in the evaluation. In this way, the DEA user is capable of changing the initial conditions in order to get entirely new results, i.e. to easily control the methodology. On the other side, it requires the solution of a mathematical program for each of the DMUs. The use of standard linear programming software may be computationally intensive in cases with large data sets.[104] A solution is offered by many specialized software tools for DEA calculation that are currently available to facilitate and accelerate the calculation process and to ensure user friendly handling.[105] Moreover, as the date used for the evaluation is provided by the DMUs themselves, it is vulnerable to manipulation. For the DEA user, it is impossible to verify all the data obtained from the DMUs. Against this background, it is even more important to carefully analyze the results of DEA concerning various aspects.

One of the most beneficial features of DEA is its capability to capture multiple relevant performance criteria, regardless of their units of measurement.[106] Thus, in contrast to other performance evaluation approaches, DEA provides the advantageous possibility of simultaneously incorporating financial as well as non-financial performance measures. This enables the DEA methodology to be adapted to new quantitative information, while qualitative information has to be transformed into quantitative one before it can be included in the evaluation. However, the flexibility of incorporating all kinds of measures into the DEA methodology bears a significant danger of mixing short-term with long-term measures as well as indices with volume measures. In order to avoid this pitfall, it is essential to carefully check the underlying data for compatibility, and – if necessary – to rescale it or to use some surrogate or rather proxy measures.[107]

---

102 Several approaches to sensitivity analysis have been proposed in the DEA literature. For a brief introduction, see Cooper et al. (2007b).

103 Cf. Charnes et al. (1994), pp. 4–5; Cook et al. (2014), p. 2.

104 Cf. Ali (1994), pp. 63–64; Charnes et al. (1994), p. 89; Ozbek et al. (2009), p. 830.

105 Cf. Paradi & Sherman (2014), p. 35.

106 Cf. Charnes et al. (1994), p. 8.

107 Cf. Dyson et al. (2001), pp. 249–251.

Since DEA is capable of simultaneously incorporate and evaluate multiple performance measures, it seems to be complete on important and relevant issues.[108] In this respect, the advantage of DEA is that it enables not only the incorporation of performance criteria under the control of DMUs, such as traditional production factors, but also additional exogenous criteria, e.g. environmental, geographical or legislative factors, which are uncontrollable by DMUs but affect their performance.[109] However, this attractive capability of DEA has also its disadvantages. Firstly, a large number of performance criteria tend to influence the evaluation of the DMUs by permitting a higher number of DMUs to be efficient. In order to avoid a low level of discrimination, some rules for the parsimonious selection of performance criteria and for the appropriate proportion of DMUs to performance criteria have been established. Secondly, DEA is based on the assumption that the performance criteria are isotonic, i.e. smaller level of inputs and larger level of outputs indicate higher level of efficiency, meaning that inputs have to be minimized and outputs to be maximized.[110] But there are exceptions to this assumption, in which inputs are to be maximized (e.g. waste determined for a waste-burning power plant) and outputs to be minimized (e.g. emissions of this power plant). Such anti-isotonic or rather undesirable performance criteria do not fit into the DEA methodology and are, therefore, difficult to handle.[111] The same applies to performance criteria that can simultaneously be characterized as an input and an output.[112] Both problems can be solved through a clear definition of the purpose and the corresponding goals of the evaluation and the subsequent connection of the performance criteria with these goals.[113] Thirdly, the endogenous determination of the weights for performance criteria allows the neglect of those criteria in which the performance of a DMU is weak by assigning (nearly) zero weights to these criteria.[114] This may distort the DEA results and is contradictory to the assumption that all of the pre-selected multiple performance criteria are included in the evaluation of performance and the evaluation is complete on important issues. In order to prevent the neglect

---

108 Cf. Epstein & Henderson (1989), p. 112; Charnes et al. (1994), p. 8.
109 Cf. Dyson et al. (2001), p. 252; Ozbek et al. (2009), pp. 829–830.
110 Cf. Dyson et al. (2001), pp. 248–250; Cook et al. (2014), p. 2.
111 Cf. Dyson et al. (2001), pp. 251–252.
112 Cf. Cook et al. (2014), p. 2.
113 Cf. Dyckhoff & Ahn (2010), pp. 1268–1269.
114 Cf. Dyson & Thanassoulis (1988), p. 564; Allen et al. (1997), pp. 14–15; Thanassoulis et al. (2004), p. 100.

of performance criteria within the evaluation, additional information can be incorporated into the analysis, e.g. constraints on weights or modifications of the data set for the DMUs.[115]

Finally, the performance reports generated by DEA can substantially assist decision makers in performance improvement.[116] Based on the calculated pareto optimal efficiency scores,[117] the identification of relative efficient as well as inefficient DMUs becomes evident. Moreover, for each inefficient DMU, DEA suggests an individual path to improve its efficiency through the detection of strong, i.e. fully, efficient DMUs as appropriate benchmarks along with the criteria-specific possible improvement potential.[118] Such comprehensive performance results are easy to communicate, especially since they enable the decision makers to conduct various "what if" scenarios to see the effects of potential decisions on the performance results of the corresponding DMUs.[119]

Although it seems quite likely to use the DEA results, e.g. efficiency scores, for a performance or rather efficiency ranking, they are not appropriate for a ranking because of the considerable differences between the reference sets, i.e. benchmarks, of the DMUs. Hence, such rankings are not meaningful and should not be established.[120] Furthermore, caution is necessary when interpreting the suggested improvement potential, as they are not necessarily attainable for the corresponding inefficient DMU. This can be explained, on the one hand, by the fact that sometimes there may be additional constraints, which restrict the ability of a DMU to produce more of the outputs, e.g. capacity restrictions, and/or to reduce the given inputs, e.g. long-term contractual relationships. On the other hand, it may be possible that a benchmark operates under specific circumstances, which are not valid for other DMUs. Therefore, the provided improvement potential should be regarded rather as a soft target, which can be adjusted to suit the circumstances.[121]

A summarizing overview of the strengths and limitations of DEA is presented in Table 2.1.

115 Cf. Thanassoulis et al. (2008), pp. 321–340.
116 Cf. Jain et al. (2011), p. 617.
117 Cf. Charnes et al. (1994), p. 8.
118 Cf. Paradi & Sherman (2014), p. 34.
119 Cf. Paradi & Sherman (2014), pp. 46–47.
120 Cf. Sinuany-Stern & Friedman (1998), p. 472; Adler et al. (2002), p. 255.
121 Cf. Cooper (2014), pp. 6–7.

Tab. 2.1: Strengths and limitations of DEA

Managerial requirements on model design[122]	**Strengths**	**Limitations**
Simple	Single aggregate measure of relative efficiency for each DMU No a priori specification of a particular production function as well as of weights for performance criteria	Demanding mathematical background, requiring assistance from DEA experts
Robust	Replicable measure of relative efficiency based on computation of best observed value	Sensitive to outliers, measurement errors, missing data, and exclusion of performance criteria
Easy to control	Focus on individual DMUs rather than population averages	Large data sets can be computationally intensive Vulnerable to manipulation
Adaptive	Different units of measurement are acceptable	Potential danger of mixing short with long term measures as well as indices with volume measures
Complete on important issues	Simultaneous consideration of multiple performance criteria Various possibilities for incorporating additional exogenous information	Large number of performance criteria diminish the discrimination of DMUs Difficulties with handling dual role and anti-isotonic (undesirable) performance criteria Permission of (nearly) zero weights, excluding respective performance criteria from the evaluation
Easy to communicate with	Identification of relative efficient and inefficient DMUs based on pareto optimal efficiency scores Identification of benchmarks and criteria-specific improvement potential	Inappropriate rankings based on the efficiency scores Suggested improvement potential not necessarily attainable

122 Cf. Little (2004), pp. 1843–1844.

2.2.4 Procedure for a DEA application

DEA models are widely applied to various organizations of public and private sector, e.g. banks, hospitals, or universities.[123] Each of these applications may rely on a multi-level modeling process that includes typical steps such as data preparation, model selection, evaluation, and interpretation of the results.[124] However, as noticed by Golany & Roll (1989), most applications in DEA provide only an insight into the underlying modeling process and not an overall view of the application procedure itself. The lack of a systematic disclosed guideline makes the traceability of the results more difficult, especially due to the fact that even slight alterations within the application procedure can lead to significantly different results.[125] Without the support of DEA experts, such as DEA researchers or scientists, practitioners are unable to cope with an appropriate application of the DEA models for their organizations.

An introduction and adherence of a general guideline for the modeling process of DEA applications would make the evaluation of efficiency comprehensible, less complicated, and more reliable. Golany & Roll (1989) provided such a general guideline in the form of a systematic flow chart with various steps. Their concept is one of the first approaches that addressed the difficulties of DEA application procedure in a methodical manner. In doing so, the authors reconciled the methodological developments with the experiences obtained from different applications. More recently, Emrouznejad & De Witte (2010) presented the "COOPER-framework" – an alternative step-by-step procedure that deals with large data samples. Though the framework consists of six interrelated phases, it resembles rather a check list for a DEA expert than an implementation guideline for an inexperienced practitioner. For this reason and due to the fact that the flow chart of Golany & Roll (1989) represents an established methodology, the following dissertation takes the procedure of Golany & Roll as a basis for DEA applications.

According to Golany & Roll (1989), there are three main phases in carrying out an evaluation of efficiency by means of DEA. The first phase aims at the

123 The most recent surveys of DEA literature are conducted by Emrouznejad et al. (2008) and Liu et al. (2013a). The former present a quantitative analysis on the whole set of DEA literature published from 1978 to 2007. The latter distinguish between methodological and application articles published from 1978 to August 2010 and focus on the most-cited methodological works for the five major DEA application areas (banking, health care, agriculture and farm, transportation, and education).

124 According to Ali (1994), p. 82, an application of DEA by means of DEA software is comprised of four steps: "data management, model selection, solution, and report generation".

125 Cf. Emrouznejad & De Witte (2010), p. 1573.

definition and selection of a homogenous population of DMUs. This task is directly related to the determination of the goals[126] of the analysis, which covers the focus of the evaluation. Depending on the decision concerning the following aspects, the focus of the analysis may vary, requiring the consideration or rather evaluation of considerably different sets of DMUs:

- *partial vs. overall evaluation,* i.e. whether a specific functional part of efficiency, e.g. cost efficiency, or the overall efficiency is evaluated;
- *internal vs. external evaluation,* i.e. whether the focus of the evaluation is within one organization or across different organizations;
- *static vs. dynamic evaluation,* i.e. whether the evaluation takes data from a single time period or from multiple periods into consideration.

Accordingly, a clear and a priori definition of the focus of the analysis enables an appropriate selection of DMUs to be evaluated. However, particular attention should be given to ensure that DMUs, which may be considered as outliers, are sifted out. Due to their deviation from the general characteristics of the considered set of DMUs, e.g. if the size of a DMU is eminently higher than the one of the remaining DMUs, such DMUs may distort the results of the evaluation.

The second phase concerns the determination of suitable performance criteria for the evaluation of the selected DMUs. To start with, all criteria that may affect the performance of DMUs – here the efficiency – should be listed. This may result in a considerably long list of relevant criteria that have to be reduced to achieve convincing results of the analysis. An unreasonably large number of performance criteria tend to influence the evaluation of the DMUs by permitting a higher number of DMUs to be efficient. In order to achieve a reasonable level of discrimination a "rule of thumb"[127] has been established, suggesting that

> [...] the number of units should be at least twice the number of inputs and outputs considered.[128]

Golany & Roll (1989) proposed three different steps for the determination of the most relevant performance criteria:

- *Judgmental screening*
  The decision makers have to examine which criterion should remain in the initial list. It has to be checked if the criteria are related to the goals of the analysis, if the information contained in the criteria is pertinent and not already included

---

126 Here, "goals" refer to the focus of the analysis and not to the goals of an organization.
127 Golany & Roll (1989), p. 239; Dyson et al. (2001), p. 248.
128 Golany & Roll (1989), p. 239.

in other criteria, and if the criteria are reliable. Particular attention should be paid to "the proper distinction between factors determining efficiency and factors explaining efficiency gaps"[129].

- *Non-DEA quantitative analysis*
  In this step, numerical values have to be attached to the criteria that are still in the list. This should be the case for quantitative as well as qualitative criteria. For the latter ones, the assignment of numerical values is essential for their participation in the performance evaluation. Depending on the goals of the analysis, it can additionally be proven if there are some criteria that can and/or should be aggregated. Another important task within this step is the classification of the criteria into inputs and outputs according to the underlying production relationships. In cases, in which the classification of some criteria is not clear, regression analyses can be carried out to indicate the tendency towards the affiliation to one group or the other.
- *DEA based analysis*
  This step aims at a close examination of criteria remaining in the list and their relation to their contribution to the analysis' results. By running trial runs of the strictest DEA model, CCR, based on different combinations of the criteria, it is possible to find out which of these criteria are really relevant and which can be left out due to scarce impact on the results.

The modifications of the initial selection of relevant performance criteria may induce a reconsideration of some of the previous steps. Especially, the number of selected DMUs has to be adjusted to the altered number of performance criteria. Hence, some steps of the procedure are applied iteratively, as illustrated by broken lines in Figure 2.7. The amount of iterations depends on each application and the preferences of the decision makers.

In the third and last phase of the procedure, different suitable DEA models can be applied to further analyze the differences in the results. For instance, the change of model orientation, i.e. input vs. output orientation, can already yield completely different results. Moreover, additional information – if available – should also be introduced into the respective DEA models. The results obtained can be analyzed concerning various aspects:

- the impact of selected criteria and respective constraints in terms of weight restrictions or managerial preferences as necessary;

129 Golany & Roll (1989), p. 241, explain the abovementioned distinction via an example of labor: "labor input may serve to determine efficiency while the scale at which the unit operates may be an explaining factor".

- special analyses, such as e.g. clustering of DMUs, partial analysis with regard to a sub-set of goals chosen for the analysis, or outlier detection by means of statistical methods;
- the composition of benchmarks, efficiency scores, and targets across different models or model orientations for each DMU.

The appropriateness and the extent of above mentioned analyses of the results depend on the specific application and should be judged by the decision makers.

Fig. 2.7: DEA application flow chart[130]

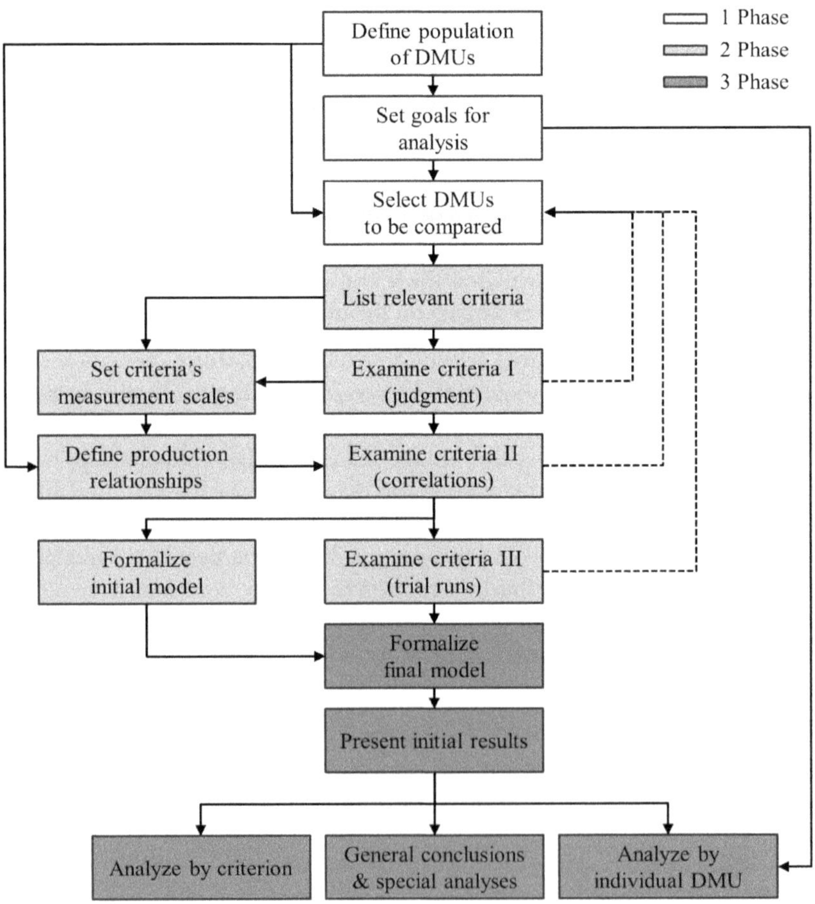

130 Cf. Golany & Roll (1989), p. 240.

Figure 2.7 illustrates the application procedure of Golany & Roll (1989). The multiple steps are highlighted by different shades of grey to emphasize their affiliation to the three main phases of the approach. The steps of the first phase are light grey, the steps of the second phase are middle grey, and the ones of the third phase are dark grey. To elucidate the potential of DEA as an analytic technique for performance management, the next section presents an applicable and comprehensive performance evaluation framework derived from the incorporation of DEA into the core element of the PMP – the performance evaluation process.

2.3 Performance evaluation based on DEA

2.3.1 Performance evaluation process (PEP)

Performance evaluation – as an essential part of performance management – is a comprehensive process carried out to gather, analyze, and provide all relevant performance information necessary to monitor the performance of an organization in relation to its goals. This includes not only the measurement of the actual level of performance, but also the more important tasks, the disclosure of performance gaps and the identification of improvement potential.[131]

According to Afsharian et al. (2014), a *performance evaluation process* (PEP) proceeds in three main stages:

- framework building,
- performance measurement, and
- performance assessment.

The first stage – framework building – provides the basis for the entire PEP. It requires an accurate understanding of organizational goals and the composition of organizational performance to determine which subunits should be evaluated to calculate their positive or negative contribution to the accomplishment of organizational goals. Such subunits could be organizational business units, departments, stores, branches as well as activities and processes, or even individuals.

Next, a set of suitable and measurable performance criteria, familiarly known as KPIs, which best capture the performance of the units under consideration, should be specified. Usually, such performance criteria focus on financial data directly available from managerial accounting systems.[132] However, the use of financial measures for performance evaluation has been criticized as being too

131 Cf. Bredrup (1995b), pp. 172–173; Bredrup (1995c), p. 191.
132 Cf. Bredrup (1995b), p. 169; Ghalayini & Noble (1996), p. 64.

"backward-looking",[133] concentrating on short-term rather than long-term aspects of managerial actions.[134] This leads to an increased emphasis on non-financial measures, such as customer satisfaction, product quality, etc., which indicate future performance and encompass all aspects of organizational performance together with the financial measures.[135] Consequently, the challenge within this stage is to provide a balanced set of performance criteria,[136] which includes both financial and non-financial measures but does not evoke information overload.[137]

Finally, the methodology that should be applied in the next stage of PEP to measure the performance of each organizational unit has to be selected. A recent review of the state of the art literature on performance measurement, provided by Nudurupati et al. (2011), mentions some of the widely recognized performance measurement systems, such as performance measurement matrix, performance prism, strategic measurement and reporting technique.[138] The choice of a measurement approach suitable for a specific application depends on the purpose of the performance evaluation, the demands of the decision makers on the methodology, and the available information.[139]

In the second stage of PEP, the performance measurement of the selected organizational units occurs by incorporating all previously specified performance criteria into the methodology currently chosen. The results – performance levels for each unit – indicate the extent to which the units succeed in attaining organizational goals. However, they do not provide the reasons for these outcomes.[140] The aim of performance measurement lies in the quantification of performance,[141] but not in its judgment.[142] It rather yields essential information for further analysis,[143] which takes place in the final stage of PEP – the performance assessment.

---

133 Ittner et al. (2003b), p. 717; Chow & Van der Stede (2006), p. 2.

134 Cf. Nudurupati et al. (2011), p. 280.

135 Cf. Ittner & Larcker (1998), p. 209; Banker et al. (2000), pp. 66–67; Ittner et al. (2003a), p. 726; Chow & Van der Stede (2006), p. 7.

136 Cf. Kaplan & Norton (1992), p. 71; Waggoner et al. (1999), p. 54; Lohman et al. (2004), p. 269.

137 Cf. Schick et al. (1990), p. 199; Ittner & Larcker (1998), p. 226; Sprinkle & Williamson (2007), p. 430; Nudurupati et al. (2011), p. 279.

138 A detailed description of the frameworks can be found in the original sources: Keegan et al. (1989), Lynch & Cross (1991), and Neely et al. (2002).

139 Cf. Bredrup (1995b), p. 189.

140 Cf. Cook et al. (1995), p. 1304; Amaratunga & Baldry (2002), p. 218.

141 Cf. Bredrup (1995b), p. 174; Neely et al. (1995), p. 80.

142 Cf. Spitzer (2007), p. 42.

143 Cf. Lohman et al. (2004), p. 271.

Performance assessment has the responsibility "to look behind the measures"[144], i.e. to interpret and analyze the results of performance measurement. Performance assessment focuses on the analysis of the cause and effect chains among different performance criteria in order to identify improvement potential as well as the reasons for performance gaps, i.e. the variances between current and desired performance.[145] Thereby, it not only assists the decision makers in understanding the impact of past decisions to monitor current (past) performance,[146] but also provides essential information that facilitate future decision making and control within the process of performance management. Figure 2.8 illustrates the 3-stage process of performance evaluation.

Fig. 2.8: Performance evaluation process[147]

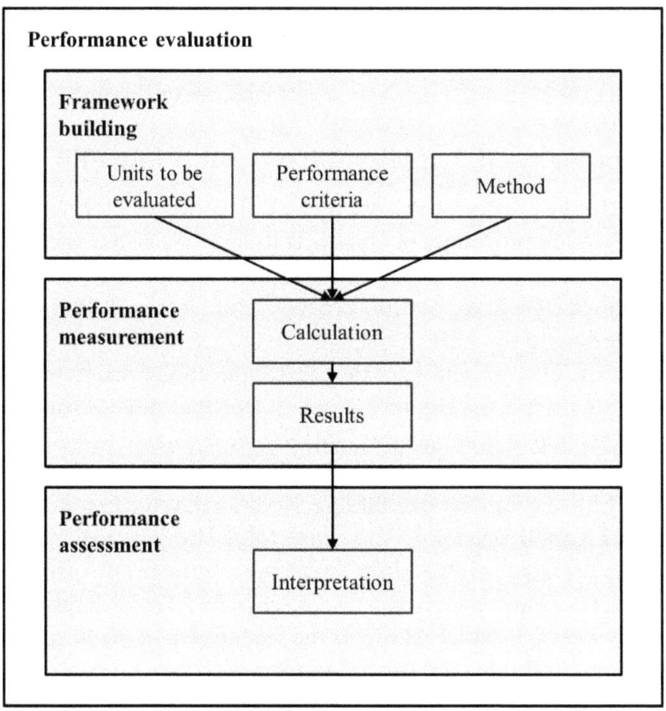

144 Bredrup (1996b), p. 191.
145 Cf. Bredrup (1996b), p. 191; Grafton et al. (2010), pp. 692–693.
146 Cf. Grafton et al. (2010), p. 693.
147 Afsharian et al. (2014), p. 255.

It is obvious, that the configuration of the PEP is of subjective nature, since the first and third stages of PEP require judgments of the decision makers.[148] However, as mentioned by Bredrup & Bredrup (1995),

> [o]bjectiveness is impossible, but a well-prepared process of subjective evaluation is sufficient […].[149]

As subjectivity cannot be avoided, the presence of it is acceptable within the PEP as long as the entire process is clear, comprehensible, and well-prepared. Here, DEA, introduced in the previous section, comes into focus. In order to support the managerial decision making within each stage of PEP and to generate reliable and meaningful results, which are of critical importance for the subsequent step of PMP – the performance control, it seems useful to integrate the DEA into the PEP. The established DEA application procedure of Golany & Roll (1989) is thereby used as a basis for a clear and comprehensible modelling process of DEA. The following subsection presents an integration of the DEA application procedure of Golany & Roll (1989) into the process of performance evaluation.

2.3.2 Integration of DEA application procedure into PEP

As mentioned in the previous subsection, the PEP proceeds in three main stages. The DEA application procedure of Golany & Roll (1989) also originally includes three main phases. However, these phases are not identical with the stages of the PEP. Therefore, the DEA application procedure of Golany & Roll (1989) is re-classified according to the PEP.

The steps of the first two phases of the DEA application procedure, highlighted by light and middle grey in Figure 2.7, aim at the definition of the focus of the analysis, the selection of homogenous DMUs to be evaluated as well as the determination of relevant performance criteria. Therefore, they clearly belong to the first stage of PEP – the framework building. Additionally, the first step of the third phase of the DEA application procedure, highlighted by dark grey in Figure 2.7, is assigned to the first stage of PEP, since it chooses a suitable DEA model required for performance measurement in the next stage of PEP. Accordingly, the calculation and presentation of initial results, as the second step of the third phase of the DEA application procedure, are part of performance measurement – the second stage of PEP. Finally, the last steps of the DEA application procedure focus on a further analysis and interpretation of the initial results, thus they obviously belong to the last stage of PEP – the performance assessment.

148 Cf. Afsharian et al. (2014), p. 255.
149 Bredrup & Bredrup (1995), p. 113.

Fig. 2.9: Integration of the DEA application procedure of Golany & Roll (1989) into PEP and assignment of identified research questions

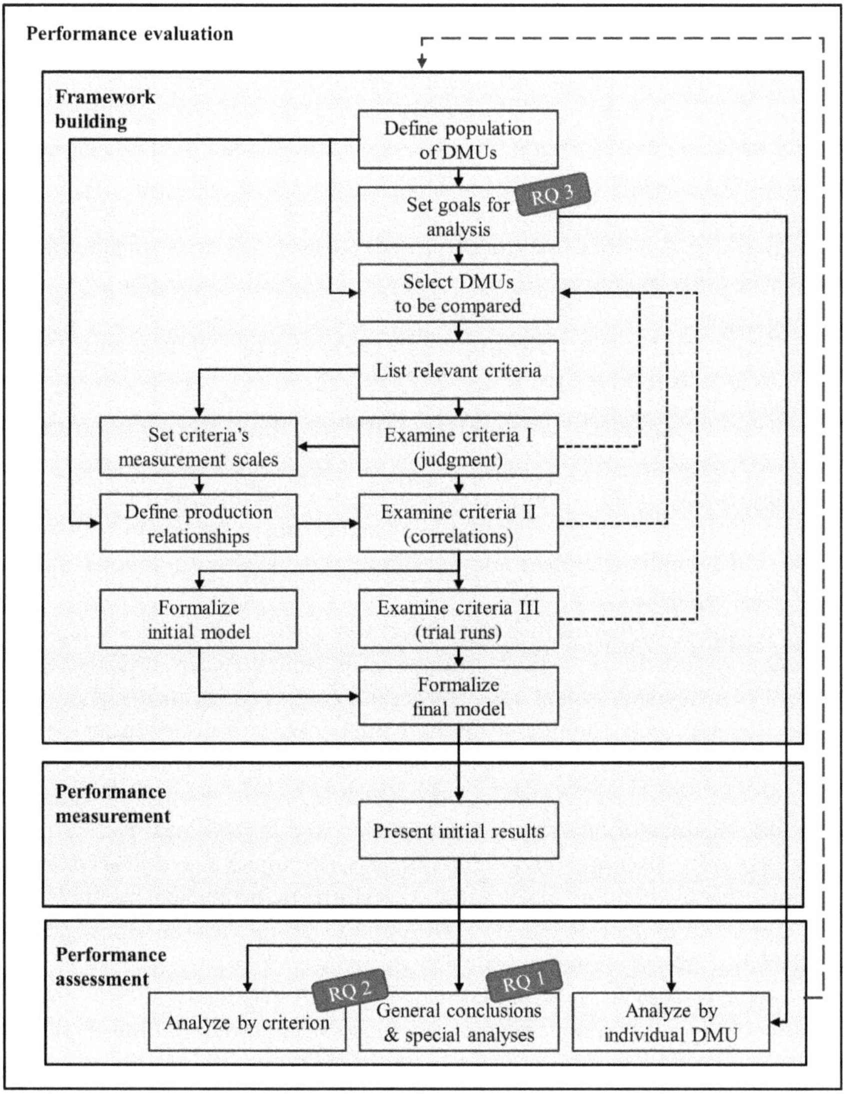

Figure 2.9 illustrates the integration of the DEA application procedure of Golany & Roll (1989) into PEP proposing to feedback information gained within the last stage into the beginning of the PEP – the framework building – as illustrated

by the broken line. This feedback loop allows, if necessary, a reconsideration of initially imposed assumptions and decisions in order to improve the modelling and measurement process with the aim of generating accurate and meaningful results.

Additionally, possible potential for methodological developments are detected in PEP, as illustrated by the three dark grey patches. The assignment of these patches is done according to the starting points of the research questions (RQ) that are addressed in detail in this dissertation. The methodological developments are motivated by the intention to meet the demands of business practice, on the one hand, and to rectify some of the limitations of DEA, on the other. The next subsection addresses these methodological developments and formulates the respective research questions.

2.3.3 Potential for methodological developments of performance evaluation based on DEA

The previously presented incorporation of DEA into the PEP provides an adaptable, comprehensive, and systematic step-by-step framework for the performance evaluation based on DEA. In order to advance the use of this framework in practice, it is analyzed for possible methodological developments of DEA. In doing so, particular emphasis is placed on the elimination of some limitations of DEA that entail managerial problems.

The potential of DEA for managerial decision making is particularly apparent in the final stage of the PEP – the performance assessment. At this stage, the results of performance measurement are analyzed in detail, to assist the decision makers, such as managers, in understanding the impact of past decisions on current performance, on the one hand, and in supporting and facilitating future decisions by suggesting ways for performance improvement, on the other. It is thus hardly surprising that practitioners are primarily interested in this stage. Based on the experience from discussions with practitioners, two topics above all seem to receive managerial attention: the different possibilities to further analyze the results of performance evaluation based on DEA and their significance with regard to multiple dimensions of organizational performance.

However, the results generally provided by DEA encompass mostly one dimension of organizational performance – the relative efficiency. Indeed, performance is a multidimensional concept. Thus, there are additional dimensions of performance that may be of managerial relevance. Effectiveness represents one of them. It usually addresses the question to what extent the fundamental goals of an organization, which are essential for its long-term business success, are fulfilled.[150] Neglecting the aspect of effectiveness within the performance evaluation based on DEA results in an incomplete picture of organizational performance.[151] Accordingly, there is considerable potential for methodological development in terms of extending DEA for the evaluation of effectiveness.[152] Given the aim of better aligning the scientific approaches with the demands of practitioners, special attention should be devoted to the requirements of business practice concerning the evaluation of effectiveness. Here, the incorporation of information about predetermined goals and corresponding target levels usually available in the context of common performance management systems, as well as the consideration and evaluation of goal achievements beyond the predetermined targets, are depicted as relevant requirements. This poses the first research question:

RQ 1: Which DEA approaches for the evaluation of effectiveness are currently available, to what extent do they meet the requirements of business practice and how could a new approach be developed to satisfy all of the depicted requirements of business practice?

Performance criteria or rather KPIs, used to capture the performance of the DMUs under consideration, provide an essential base for the performance evaluation by means of DEA. Therefore, their impact on the results of the evaluation constitutes a further aspect of interest. Due to the need of parsimonious selection of performance criteria, solely the most important and meaningful measures are included in the evaluation. Accordingly, it is assumed that all of them are taken into account within the evaluation and calculation of the results. However, the flexibility in weight determination, as one of the key strengths of DEA and simultaneously one of its shortcomings, allows neglecting those criteria in which the performance of a DMU is insufficient. This may cause considerable managerial problems, such as misallocation of resources by concentrating on only some of the crucial performance criteria.[153] Such extremism in criteria achievement is contradictory to the abovementioned assumption that attention is paid to all selected performance criteria and may result in a questionable evaluation of organizational performance. Therefore, the idea arises to consider balance in the

---

150 A deeper discussion of the concept of effectiveness is content of Chapter 3.

151 Cf. Kumar & Gulati (2010), p. 52.

152 The consideration of additional dimensions of performance may be part of special analysis within the stage of performance assessment. This corresponds to the assignment of the first research question in Figure 2.9.

153 A deeper discussion of problems resulting from flexible weights is content of Chapter 4.

criteria achievement of the DMUs as an additional dimension of performance, which should be able to provide important insight into a DMU's performance structure. As this dimension is so far not regarded within the performance evaluation, there is, consequently, further potential for methodological development not only concerning the exclusion of performance criteria – resulting from flexible weights – but also the extension of DEA for the evaluation of balance.[154] This gives rise to the second research question:

RQ 2: *Which DEA approaches that address the possible neglect of performance criteria due to the flexibility in weight determination are currently available and how could a new DEA approach be developed for the evaluation of balance?*

It is clear that performance criteria play an important role in the performance evaluation by means of DEA. The selection of suitable criteria and their subsequent characterization in the inputs or outputs represent, however, two challenging tasks. The extensive selection process in the first stage of the framework in Figure 2.9 confirms this fact. By means of three different steps, the most relevant performance criteria are determined. Although, it is presupposed that the criteria are related to the goals of an organization, a systematic link between goals and criteria is not specified at all. This hinders the transparency of the criteria selection, undermining the plausibility of the results, since it is unclear if all organizational goals are completely covered by the criteria selected for the evaluation. As a consequence, the subsequent characterization of the selected performance criteria as inputs or outputs is also questioned, especially if some of the criteria are anti-isotonic and/or have a dual form, i.e. are inputs and outputs at the same time. Such criteria do not fit into the traditional DEA methodology and their consideration so far is not possible without quality impairment of the results. Both of these pitfalls disclose the conceptual weaknesses in the first stage of the framework presented in Figure 2.9 with severe consequences not only for the entire performance evaluation, but for performance management in general.

As the alignment of organizational performance with the strategic goals of this organization is decisive for the continued existence of the organization, it seems reasonable to expand the traditional DEA methodology by a generalized DEA (GDEA). GDEA is capable of explicitly incorporate organizational goals as

154 As the second research question concerns two aspects of the performance assessment, the implicit analysis of the impact of the selected performance criteria on the results and the extension to special analysis with regard to different performance dimensions, it is assigned to both of the corresponding steps in Figure 2.9.

fundamental objectives into the evaluation, operationalize them, and connect with corresponding performance criteria.[155] Thus, it is obvious that the framework in Figure 2.9 has significant potential for methodological development concerning the systematic link between organizational goals and suitable performance criteria required to their operationalization. This also concerns the selection of performance criteria and their sometimes difficult characterization as inputs and outputs, which is no longer necessary in GDEA.[156] This leads to the third research question:

RQ 3: How is the traditional DEA extended by GDEA to ensure that performance criteria are explicitly and systematically linked to the organizational goals and how should the respective goal-oriented DEA approach for the determination of performance criteria proceed?

As already mentioned in Subsection 2.2.3, there are further limitations of DEA, but it would go beyond the scope of this dissertation to provide a solution to all of these problems. Instead, only those are addressed in details which seem to be of managerial interest and may increase awareness and use of the proposed framework by practitioners.

2.4 Conclusion and discussion

This chapter has addressed the conceptual problem of defining a comprehensive and systematic framework for performance evaluation based on DEA. Therefore, performance evaluation has firstly been delineated from performance measurement to clarify their roles within the integrated concept of performance management, which has been defined as a continuous process intended to support an organization in monitoring its performance in relation to its goals. Performance evaluation has been detected as an essential element of this process. It represents a comprehensive process carried out to gather, analyze, and provide all relevant performance information.

However, the performance evaluation of modern organizations is proved to be complex due to their numerous performance measures. Therefore, DEA as an analytic technique capable of resolving this problem has been introduced and

155 GDEA originates from Dyckhoff & Allen (2001) in the context of measuring ecological efficiency and is described in detail by Dyckhoff & Ahn (2010).

156 The assignment of the third and last research question in Figure 2.9 corresponds to the starting point of the methodological improvement potential, which will obviously affect all of the subsequent steps within the first stage of the proposed framework.

subsequently incorporated into the process of performance evaluation. This has resulted in an adaptable, comprehensive, and systematic step-by-step framework for the performance evaluation based on DEA. To prevent this framework from being ignored by practitioners, potential for methodological developments of managerial relevance have been disclosed in order to be addressed in this dissertation:

- Besides the evaluation of efficiency, additional dimensions of performance should be considered in the framework. This may be part of special analyses that take place in the last stage of the framework – the performance assessment. In addition to efficiency, effectiveness is regarded as the most well-known and meaningful performance dimension. Therefore, the first methodological development consists in extending the DEA for the evaluation of effectiveness by incorporating the requirements of business practice depicted as relevant within this dissertation.
- The impact of the selected performance criteria on the results of the evaluation constitutes a further aspect of interest within the performance assessment stage. The possibility of neglecting some of the performance criteria within the evaluation is considered as contradictory and may cause considerable managerial problems, inter alia, extremism in criteria achievement. Therefore, the idea arises to consider balance in the criteria achievement of the DMUs as an additional dimension of performance. Accordingly, the second methodological development consists in preventing the exclusion of performance criteria, resulting from flexible weights, and extending the DEA for the evaluation of balance.
- The monitoring of organizational performance with respect to its goals requires the consideration of them within the evaluation of performance. However, the extensive selection process of performance criteria that takes place within the first stage of the framework so far does not specify a systematic link between goals and criteria. This impedes the inclusion of anti-isotonic and dual criteria and undermines the plausibility of the results. Broadening the methodology of the DEA by GDEA allows the explicit incorporation of organizational goals and their linkage with suitable performance criteria, which can be of isotonic, anti-isotonic, and/or dual nature. Therefore, the third and last methodological development consists in proposing a goal-oriented procedure for the determination of performance criteria according to GDEA.

Each of the following three chapters addresses one of the three methodological developments detected within the proposed conceptual framework of performance evaluation based on DEA.

3 Evaluation of relative effectiveness[157]

A great part of the management literature as well as the majority of organizations focuses on performance management as optimization of efficiency, thereby disregarding the aspect of *effectiveness*.[158] This neglects the fact that it is of vital importance for long-term business success to monitor the fundamental goals of an organization.[159] The question to what extent such goals are achieved is usually addressed by the concept of effectiveness. In contrast to efficiency, which has been recognized as a component of performance measurement and management for a long time, effectiveness has only recently begun to receive considerable attention.

In the field of organizational theory, effectiveness has already been a central topic. The wide scope of the literature on the issue of organizational effectiveness documents this fact, by simultaneously revealing the disagreement on the concept of effectiveness. Moreover, the diversity of approaches makes it difficult to establish a comprehensive way for the evaluation of effectiveness. Based on an overview of the current theoretical approaches and problems in the context of effectiveness evaluation, the aim of this chapter is to introduce a new approach that incorporates information provided by performance management systems in terms of predetermined goals and corresponding target levels. This approach should meet the requirements on model design from the managerial point of view in order to be used by practitioners.

This chapter is organized as follows. Section 3.1 discusses the conceptual disorder in the pertinent literature by considering the evolution of organizational effectiveness and the major theoretical approaches. Based on the described findings, a definition of effectiveness is provided that underlies this dissertation. Section 3.2 describes the most substantial problems concerning the evaluation of effectiveness and recommendations provided in the literature as solutions to these problems, leading to the consideration of DEA as a useful approach for evaluating effectiveness. Therefore, a brief review of current contributions regarding the

---

157 The content described in this chapter has been partially published by Ahn, H. & Neumann, L. (2014): Measuring effectiveness – a DEA approach under predetermined targets, in: International Journal of Business Analytics 1(1), pp. 16–28. Copyright 2014, IGI Global, www.igi-global.com. Reused by permission of the publisher. The content contains some minor changes in comparison with the published version.

158 Cf. Mouzas (2006), pp. 1124–1125.

159 Cf. Ebnerasoul et al. (2009), p. 44.

evaluation of relative effectiveness by means of DEA is provided. Complementing theoretical recommendations with the requirements of business practice leads to the critical discussion of DEA literature previously reviewed. Since none of the presented approaches is able to meet all depicted requirements of business practice, a new DEA-based approach is presented in Section 3.3. The numerical example illustrates the approach and provides the basis for managerial implications explained in Section 3.4. Finally, Section 3.5 concludes the chapter and discusses the characteristics of the introduced approach.

3.1 Effectiveness of organizations: putting the conceptual disarray in order

3.1.1 Evolution of organizational effectiveness

Organizational effectiveness (OE) constitutes one of the most extensively researched topics in the field of organizational theory.[160] The interest in the effectiveness of organizations has been continuously increasing since the 1950s. A broad range of researchers has focused on the conceptualization, measurement and assessment of OE. Especially in the period between the 1970s and the 1980s, there has been a fairly intensive exploration of the theoretical construct of OE that resulted in numerous studies, articles, and monographs.[161] Despite the considerable attention given to this construct, a commonly accepted definition of effectiveness is still missing.[162] Instead, the relevant literature is characterized as noncumulative, ambiguous, and even full of controversy.[163]

The main reason responsible for the conceptual disorder in the pertinent literature concerns "the lack of success in delineating the boundaries of OE"[164]. Since the meaning of OE is closely related to the characteristics of the respective organization,[165] OE has been interpreted in a variety of ways leading to multiple approaches that will be discussed in more detail in the next subsection. The meaning of effectiveness alters thereby from one approach to the other, depending on the purpose of the investigation and the viewpoint that is being taken, e.g. the perspective of shareholders, managers or employees of the organiza-

160 Cf. Rojas (2000), p. 97.
161 Cf. Cameron & Whetten (1983a), p. 1.
162 Cf. Henri (2004), p. 97.
163 Cf. Lee & Brower (2006), p. 155.
164 Cameron & Whetten (1983a), pp. 8–9.
165 Cf. Goodman & Pennings (1979), p. 3; Cameron & Whetten (1983a), p. 4.

tion.[166] Consequently, (even entirely) different sets of criteria were considered as appropriate to evaluate effectiveness. The choice of suitable criteria relies thereby on individual preferences and judgments.[167] The resulting diversity of the criteria across the literature prohibits the development of consensual and valid criteria of effectiveness as well as a common concept of OE.[168]

Consistent with the preceding findings, Cameron (1986) summarized that the

> [c]onclusions about the consensus and conflict that surround effectiveness lead to the realization that agreement about effectiveness is mainly an agreement to disagree.[169]

More recently, Lee & Brower (2006) even declared that

> [g]iven the breadth of scope of interest of scholars and practitioners, we submit that it is unlikely that consensus will be achieved any time soon regarding definitions, a set of conventional indicators, or a common theory of organizational effectiveness.[170]

Cameron & Whetten (1983b), however, pointed out that although there is no single universal model of OE, it is neither required nor useful. Instead, they argued that multiple approaches of OE are all equally valid, but in different contexts and for different kinds of organizations. Moreover, they recommended not to establish additional theories of OE, but to develop frameworks for the evaluation of effectiveness with the aim of stimulating the understanding of OE. To allow the comparison of such frameworks, Cameron & Whetten (1983b) provided seven highly interrelated decision guidelines in form of questions that should be applied and answered in every investigation of OE. In the following, these guidelines are briefly presented.[171]

1. *From whose perspective is effectiveness being judged?*
   As explained above, the definition and evaluation of effectiveness differ significantly across the various approaches depending on the viewpoint that is being taken. Thus, there is no best perspective from which effectiveness should be assessed. Nevertheless, it is important that the specific point of view is concretized to comprehend the results of the underlying investigation of effectiveness.

---

166 Cf. Campbell (1979), pp. 16–18; Goodman & Pennings (1979), pp. 3–5; Cameron & Whetten (1983b), p. 263.

167 Cf. Campbell (1979), pp. 15–18; Scott (1979), pp. 72–73; Cameron & Whetten (1983b), p. 264.

168 Cf. Cameron & Whetten (1983b), p. 265.

169 Cameron (1986), p. 544.

170 Lee & Brower (2006), p. 156.

171 For an elaboration, see Cameron (1980), Cameron & Whetten (1983b), and Cummings (1983).

2. *On what domains of activity is the judgment focused?*
   Since organizations contain a variety of domains, it is enormously difficult for the organizations to be maximally effective in all of them. Furthermore, the importance of the domains can change over time, meaning that evaluating the effectiveness of an inappropriate domain would be misleading. To avoid inaccurate judgments of effectiveness, the domain or domains being evaluated should be clearly specified.
3. *What level of analysis is being used?*
   Evaluation of effectiveness can be carried out at different levels of an organization, e.g. evaluation of the effectiveness of individuals, subunits or the whole organization. Since the effectiveness of these levels is not always compatible, it is crucial to the evaluation and the expecting results that the most appropriate level is selected. Obviously, the decision should be made in accordance with the perspective and the domain or domains being chosen as well as the purpose of the evaluation.
4. *What is the purpose of judging effectiveness?*
   The decision about the purpose for the evaluation of effectiveness is of immense importance, because it affects the evaluation itself by generating completely different results and consequences for the underlying unit. Therefore, it is required that a clear definition of the purpose is determined a priori. Moreover, the purpose of the evaluation is interrelated to the other six questions and cannot be considered in isolation and vice versa.
5. *What time frame is being employed?*
   Different organizations have different strategies concerning effectiveness, e.g. some focus only on achieving short-term effectiveness, while others surrender short-term effectiveness in favor of long-term effectiveness. The effects of these strategies occur in different time frames as well. Hence, not to consider the time frame will impede convincing judgments of effectiveness. Therefore, it is necessary to take a closer look at the time frame in order to select the appropriate one for the evaluation.
6. *What type of data is being used for judgments of effectiveness?*
   The evaluation of effectiveness is impossible without data. Data can be obtained from various sources, e.g. collected by the organization itself (internal data) or based on the perceptions of different external experts (external data), it can contain diverse information, and have different characteristics, e.g. being quantifiable or not, objective or subjective, etc. Therefore, data used for the evaluation of effectiveness should be carefully selected in consistence with the

preceding decisions concerning the purpose, level of analysis, time frame and so on.

7. *What is the reference against which effectiveness is judged?*
   In general, effectiveness is determined in relative terms, i.e. the actual result is compared to a reference, as a standard. The degree to which the actual result approximates the reference represents a measure of relative effectiveness. The necessary condition is that the reference is available in operational form.[172] Due to the dependency of the effectiveness evaluation on the underlying reference and the fact that various references could be chosen, it is advisable to be clear about which reference to select for the specific evaluation of effectiveness.

3.1.2 Major theoretical approaches to organizational effectiveness

This subsection reviews four major approaches to organizational effectiveness, which have been identified as fundamental for the research and the subsequent development of modifications and alternative approaches. None of these approaches is superior to the others, because each of them represents a different view concerning the definition and the evaluation of effectiveness. In addition, they reflect the chronological evolution of the theoretical conceptualization of OE as presented in Figure 3.1.

Fig. 3.1: Four major theoretical approaches to organizational effectiveness

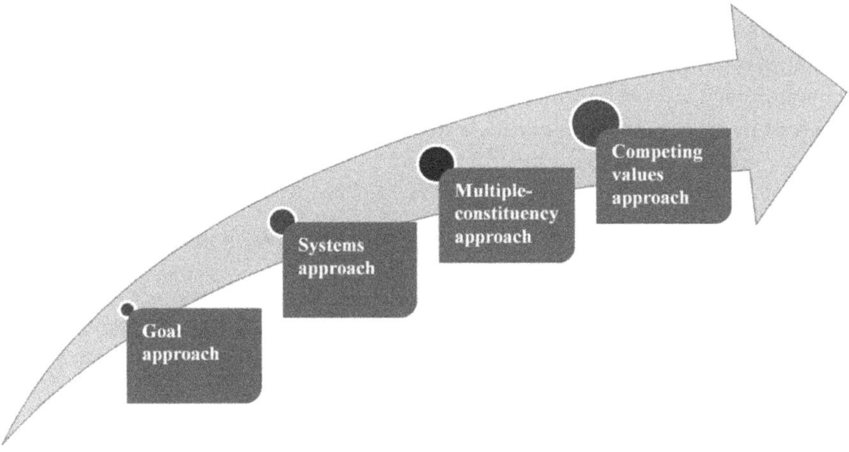

172 Cf. Pennings & Goodman (1979), pp. 161–162.

The earliest and most widely used approach arises from the assumption that the concept of effectiveness is inevitably related to the concept of organizational goals.[173] Goals serve as standards by which the success of an organization – and therefore the effectiveness – could be evaluated.[174] The *goal approach* seizes on this finding and defines effectiveness as the extent to which an organization accomplishes its stated goals.[175] Obviously, the goals should be unambiguous, measurable, and manageable.[176] Implying that an organization usually contains various goals, a judgment is required concerning the selection of the goals relevant for the evaluation.[177] As a consequence, outcomes of the evaluation of effectiveness are certainly influenced by such judgments.[178] A further shortcoming of the goal approach appears in connection with the determination of references against which the organizational goals are compared. As Etzioni (1960) pointed out, if the references are not really attainable or too low, the results of the effectiveness evaluation will always be misleading.

The dissatisfaction with the goal approach resulted in the development of a new concept – the *systems approach*.[179] Whereas the goal approach regards an organization as a rational system allowing some interest groups to accomplish certain goals, the systems approach defines it as a social unit that not only strives for the achievement of some sustainable goals but simultaneously carries out necessary activities for its maintenance.[180] The organization, therefore, focuses mainly on the study of its internal processes – the acquisition and distribution of resources, their transformation in outcomes, etc. – to preserve its integration and adaptation in the environment.[181] Effectiveness is evaluated by this approach as the extent to which an organization acquires scarce resources from its environment.[182] The systems approach is most fruitful if there is a clear connection between the resources received by an organization and its outcomes, e.g. a bank is more profitable, if it

---

173 Cf. Hannan & Freeman (1979), p. 110; Scott (1979), p. 64; Rojas (2000), p. 98.

174 Cf. Etzioni (1960), p. 257.

175 Cf. Etzioni (1960), p. 257; Cameron (1980), p. 67; Seashore (1983), p. 58; Au (1996), p. 2; Forbes (1998), pp. 185–186; Rojas (2000), p. 98; Lee & Brower (2006), p. 158.

176 Cf. Campbell (1979), p. 19; Cameron (1980), p. 68; Seashore (1983), p. 58; Forbes (1998), p. 185.

177 Cf. Campbell (1979), p. 15.

178 Cf. Etzioni (1960), p. 258; Lee & Brower (2006), p. 159.

179 Cf. Rojas (2000), p. 98; Zammuto (1984), p. 606.

180 Cf. Etzioni (1960), p. 261; Pennings & Goodman (1979), p. 148.

181 Cf. Campbell (1979), p. 20; Scott (1979), p. 74.

182 Cf. Cameron (1980), p. 67; Forbes (1998), p. 186.

accepts more deposits and makes more loans.[183] Despite the provision of a more comprehensive view of the organization, the systems approach was criticized for placing more emphasis on non-rational goals as the rational ones.[184] Furthermore, the complexity of internal system relationships required for this approach makes the evaluation of the effectiveness more difficult.[185]

In an attempt to mitigate the shortcoming of the goal and systems approaches, a *multiple-constituency approach* emerged that extends the scope of both previous approaches by treating effectiveness as a set of multiple different statements. These statements reflect the relevant goals determined by the multiple constituencies[186] associated with the particular organization. In contrary to the previous approaches, the multiple-constituency approach does not assume that there is a single unambiguous statement of the organization's purpose. Rather, it recognizes the differences in the demands of varied constituencies on the outcomes of the organizational performance.[187] Therefore, multiple evaluations of effectiveness arise from the perspectives of different constituencies. For example, it is possible that an organization could be simultaneously assessed as effective by its management and ineffective by the government.[188]

The evaluation of effectiveness becomes complex, if the multiple constituencies of an organization focus on contradictory goals, goals in different dimensions and conflicting time frames.[189] Therefore, the importance of values of different effectiveness goals increases.[190] Based on these values and the fact that goals of multiple constituencies might simultaneously take opposite directions, the *competing values approach* emerged. This approach proposes three value dimensions that underlie the concept of OE. The first dimension refers to the organizational focus, i.e. from an internal emphasis on the developments within the organization to an external emphasis on the development of the organization itself. The second dimension refers to the organizational structure, i.e. from a prioritization

183 Cf. Cameron (1980), p. 68.
184 Cf. Au (1996), p. 3.
185 Cf. Au (1996), p. 3; Ahn, (1997), p. 25.
186 According to Connolly et al. (1980), p. 213, constituencies represent different individuals and groups that are directly or indirectly associated with an organization, e.g. employees, managers, suppliers, customers, government agencies, and environmental groups.
187 Cf. Connolly et al. (1980), pp. 212–216; Zammuto (1984), pp. 606–614.
188 Cf. Connolly et al. (1980), p. 212.
189 Cf. Chelladurai (1987), p. 42.
190 Cf. Zammuto (1984), p. 614.

on stability to a prioritization on flexibility. The third dimension relates to organizational means and ends. These competing value dimensions are juxtaposed and lead to four different models of OE: rational goal model, internal process model, open system model, and human relations model.[191]

The competing values approach encompassed for the first time the aspects of different basic theoretical approaches to OE in a single conceptual framework. Each model contributes to the explanation of the organization as a construct. Although some of these models are at opposite locations concerning the three value dimensions, they are not necessarily mutually exclusive. This emphasizes that the recommendations derived from this approach should take into account the possible contradictions in every organizational setting.[192]

The belief that effectiveness goals and/or criteria are rather competing than congruent provides an additional explanation for such diversity in effectiveness conceptions and evaluation approaches. Cameron (1986) confirmed this perception by detecting effectiveness as a paradox:

> [o]rganizational effectiveness is inherently paradoxical. To be effective, an organization must possess attributes that are simultaneously contradictory, even mutually exclusive.[193]

The assumption of contradictory goals and/or criteria coincides with the experiences in practice. Therefore, the consideration of paradoxical effectiveness criteria within the evaluation of effectiveness gains on recognition. To manage the incorporation of such goals and/or criteria and still enhance effectiveness, Cameron (1986) suggested focusing on balance in these opposing goals and/or criteria. This recommendation will be taken into account more precisely in Chapter 4. Based on the goal-oriented preferences as essential parts of the presented approaches to OE a delineation of effectiveness from the efficiency concept is presented in the next subsection.

3.1.3 Effectiveness vs. efficiency

Although effectiveness and efficiency are among the most frequently used terms of economic theory and constitute one of the major themes of management control, their differentiation is often vague.[194] On the one hand, several even mutually contradictory interpretations across the branches of science exist.[195] On the other

191 Cf. Quinn & Rohrbaugh (1983), pp. 369–370.
192 Cf. Quinn & Rohrbaugh (1983), pp. 374–375.
193 Cameron (1986), pp. 544–545.
194 Cf. Mensah et al. (2008), p. 326.
195 Cf. Simonen et al. (2011), p. 800.

hand, the two terms are often used synonymously. Therefore, it is necessary to shed some light on the issue and give a detailed specification of effectiveness and efficiency.

The four major approaches to OE briefly described in the previous subsection show that goals are an important part in each evaluation of effectiveness. Beginning with the earliest and most widely used goal approach, goals are present and taken into account in all of the three subsequent approaches, e.g. sustainable goals within the systems approach, goals of various organizational constituencies within the multiple-constituency approach, and contradictory goals within the competing values approach. Obviously, goals seem to be adequate standards for helping to define and evaluate the effectiveness of an organization. The combination of this finding with the generally accepted concept of purpose-rationality provided by Weber (1978) offers a possibility to clearly define and differentiate between effectiveness and efficiency.

Accordingly, it will be assumed throughout this dissertation that these two terms – effectiveness and efficiency – describe aspects of purpose-rational behavior, which is characterized – among others – by the reflection on purposes and means.[196] Purposes describe the original motives pursued by an organization and its economic activities respectively, the instantiations of such purposes are called goals (in multi-criteria decision making approaches, they are characterized as fundamental goals or objectives[197]). Means are precious resources whose consumption is necessary to pursue the designated goals. This is the basis to characterize the effectiveness and efficiency as different aspects of performance. Effectiveness criteria refer to the degree to which the goals are achieved, while efficiency criteria refer to the ratio between necessary resources and the achievement of the goals.[198] These considerations are combined in the following definitions originating from Ahn & Dyckhoff (2004):[199]

> *Efficiency*: An activity is efficient with respect to a subset of the relevant goals and activity alternatives, when it leads to a change in state which – by choosing another specific alternative – allows no improvement with regard to none of the particular goals without worsening another particular goal at the same time (partial efficiency). An activity is totally efficient, when it is efficient with respect to all relevant goals and all alternatives available in the decision context.

---

196 Cf. Weber (1978), p. 26; Rutgers & Schreurs (2006), p. 406.

197 Cf. Keeney (1992), p. 34.

198 Cf. Hannan & Freeman (1979), p. 110; Athanassopoulos (1998), p. 175; Mensah et al. (2008), p. 325; Amiri et al. (2010), p. 510.

199 Cf. Ahn & Dyckhoff (2004), p. 519.

Effectiveness: An activity is effective with respect to a subset of the relevant purposes, if it leads to a change in state with which these purposes are fulfilled (partial effectiveness). It is (totally) effective, if all relevant purposes are fulfilled.

The difference between the two dimensions of performance is reflected in the more familiar definition of effectiveness as "doing the right things" and efficiency as "doing things right" which are referred to by numerous authors.[200] Obviously, effective activities are not necessarily efficient. Conversely, an efficient activity might be ineffective if the means are employed so sparingly that at least some original purposes and derived goals cannot be achieved on a satisfying level.[201]

3.2 Evaluation of effectiveness

3.2.1 Problems in evaluating effectiveness

As has been shown in the previous section, the ambiguity and confusion surrounding the concept of effectiveness have led to the development of various approaches for different kinds of organizations in different contexts. This diversity of approaches and corresponding criteria makes it difficult to establish a comprehensive way for the evaluation of effectiveness. The seven interrelated questions provided by Cameron & Whetten (1983b) offer a possibility of systematically guide and limit the scope of the investigation to ensure a valid and comparable evaluation of effectiveness. However, several problems are still inherent in evaluating effectiveness and detract from the meaningfulness of the results.

The most substantial problem arises from the inevitable involvement of decision makers in the preparation of effectiveness evaluations. According to Cameron & Whetten (1983b), there are at least seven issues – concerning the perspective, domain, level of analysis, purpose, time frame, type of data, and reference – which require judgments of the decision makers. Such judgments usually refer to individual interests, preferences and opinions. Thus, the inclusion of *subjectivity* in the effectiveness evaluation cannot be avoided. Evaluations based on subjective decisions have to accept criticism, especially regarding the possibility of distortion of the results.[202] Depending on which interests and judg-

---

200 Cf. Golany et al. (1990), p. 90; Chang et al. (1995), p. 53; Sheth & Sisodia (2002), p. 352; Asmild et al. (2007), p. 306; Amiri et al. (2010), p. 510; Lu & Hung (2011), p. 1351.

201 Cf. Drucker (1963), p. 54.

202 Cf. Scott (1979), p. 89.

ments are included, the evaluation of effectiveness might be variously influenced leading to considerably different results and implications.[203] As a consequence, organizations under consideration might be less likely to accept and trust the results of an analysis of their effectiveness.

Another considerable problem concerns the *availability and incompleteness of data*. Data about the criteria and measures of effectiveness serves as a necessary basis for every evaluation. But at the same time it represents one of the major impediments to the evaluations, because such data has to be collected. Although organizations are continuously improving their information systems to obtain and collect more data, the assumption that all required data will be really available and complete, is an illusion. On the one hand, the lack of care within the data collection is responsible for the occurrence of missing values or errors in several criteria. As a consequence, these criteria are often excluded from the consideration. On the other hand, the quantification of intangible qualitative criteria is often difficult. Thus, easily quantified criteria are usually preferred and applied while intangible qualitative criteria tend to be neglected. Such inconsistency and incompleteness of data obviously result in less accurate evaluations.[204]

The multidimensional nature of effectiveness entails a further difficulty for the evaluation of effectiveness – the *multiplicity of criteria and their aggregation*. Multiple criteria are necessary to encompass all aspects of effectiveness. However, not all of these criteria might have the same units of measurement, e.g. turnover in euro is incomparable to production volume in number of pieces. In such cases, the evaluation of effectiveness becomes complex, especially if some of the criteria are mutually exclusive.[205] Depending on the approach that is used, an aggregation of the criteria, e.g. by means of weights, might be useful. This in turn implies the need for judgment leading to the above mentioned problem of subjectivity.

In sum, this brief review reveals the most important aspects that impede not only the agreement on a common comprehensive and accurate framework for the evaluation of effectiveness, but also the general acceptance of such a framework in practice. The next subsection addresses some recommendations provided in the literature regarding the above mentioned issues.

---

203 Cf. Campbell (1979), pp. 15–16; Hamilton & Chervany (1981a), pp. 61–62; Cameron (1986), p. 541; Au (1996), p. 4.

204 Cf. Steers (1975), p. 553; Hamilton & Chervany (1981a), p. 61.

205 Cf. Steers (1975), p. 553; Lewin & Minton (1986), pp. 528–529.

3.2.2 Recommendations for evaluating effectiveness

The literature on the evaluation of effectiveness is rich in suggestions for improvement. Despite the tendency of the studies on effectiveness to compete with each other in order to develop a better way of evaluating, there is some agreement on the recommendations concerning the problems mentioned in the previous subsection. These recommendations are concisely presented in the following.

The lack of transparency within subjective judgments is an undesirable characteristic of almost every evaluation because the direct or indirect influence on the results proved to be elusive. The total elimination of subjectivity is not possible – as long as judgments are based on human's decisions – and even not always desirable. The aim is therefore to bring transparency into the process of subjective judgments. This could be fulfilled by considering different evaluator and/or constituent viewpoints within the determination of the purpose, goals, criteria as well as weights of the criteria necessary for the evaluation.[206] Subsequently, the effectiveness should be evaluated several times based on different assumptions resulted from various viewpoints. A rigorous comparison of the results across these evaluations might point out the respective effects of the specific subjective assumptions on the results. Additionally, such participative evaluations of effectiveness might facilitate the comprehension and communication of the results, leading to greater acceptance.[207]

The determination of adequate effectiveness goals and/or criteria could be supported by the use of strategic and operative goals already applied by the corresponding organization.[208] One the one hand, these goals provide a realistic basis for the evaluation of effectiveness, because they not only represent authorized goals but also reflect the activities within the organization in order to pursue these goals. On the other hand, it is more likely that data concerning the operative goals will be available. Nevertheless, intangible qualitative criteria should not be ignored. Indeed, there is a need to establish approaches for their quantification in order to enlarge the existing data about the performance of an organization.[209]

However, the resulting multiplicity of criteria makes the calculation of effectiveness more difficult. An approach is required that is able to handle the various criteria, which might have different units of measurement, be quantitative or

206 Cf. Steers (1975), p. 555; Hamilton & Chervany (1981a), pp. 63–64.
207 Cf. Hamilton & Chervany (1981b), pp. 83–84.
208 Cf. Steers (1975), p. 555; Hamilton & Chervany (1981a), p. 63; Mendelow (1983), p. 73.
209 Cf. Hamilton & Chervany (1981a), p. 63; Lewin & Minton (1986), p. 529.

qualitative, as well as be contradictory. Mathematical optimization approaches are recognized as a possible solution to this problem. Their most prominent representatives – linear optimization and multi-criteria optimization approaches – appear as eminently useful because of their capability to simultaneously account for multiple weighted effectiveness criteria as well as for various constraints within the organization regarding the achievement of effectiveness goals and/or criteria.[210] Moreover, such approaches allow to relatively easily examine the above mentioned effects in the effectiveness results regarding different choice of goals, criteria, and weights by means of computer simulations. Among the group of linear optimization approaches, DEA gains increasing attention in the context of effectiveness evaluation.[211]

DEA has been initially developed for the evaluation of relative efficiency of DMUs. However, its capabilities of aggregating multiple criteria to a single summary measure, the relative comparison of homogenous DMUs, and the identification of best practices have not remained unrecognized for the evaluation of effectiveness.[212] As a consequence, a new direction of research evolves from the detection that DEA can be exploited to evaluate effectiveness of DMUs.[213]

Concerning the depicted recommendations, it seems to be clear that a potentially useful approach for evaluating effectiveness could originate from the area of DEA. Therefore, a review of the existing contributions regarding the evaluation of effectiveness by means of DEA is provided in the next subsection.

3.2.3 Evaluating relative effectiveness by means of DEA: a brief review

While DEA efficiency is well-defined,[214] there is no consistently used definition of DEA effectiveness and no consensus on how DEA effectiveness should be evaluated.[215] A literature review on the key phrase "effectiveness evaluation by means of DEA" confirms that effectiveness is interpreted in a variety of ways leading to multiple DEA approaches. In accordance with the definition of effectiveness, the current scientific contributions can be classified into three categories.

---

210 Cf. Steers (1975), p. 556.
211 Cf. Lewin & Minton (1986), pp. 529–532.
212 Cf. Lewin & Minton (1986), pp. 529–530; Lee & Brower (2006), p. 162.
213 Cf. Lee & Brower (2006), p. 162.
214 Cf. Cooper et al. (2007b), p. 1.
215 Cf. Lee (2006), p. 139; Asmild et al. (2007), p. 306; Amiri et al. (2010), p. 510.

In the first category, effectiveness is associated with efficiency and is therefore usually evaluated with basic DEA models, e.g. CCR and/or BCC models. Although new evaluation approaches emerge within this category, e.g. the incorporation of weight restrictions into DEA, the combination of DEA and bootstrapping, these approaches still refer to effectiveness as a special kind of efficiency. Figure 3.2 illustrates the first category of the reviewed literature.

Fig. 3.2: First category of the reviewed DEA-literature

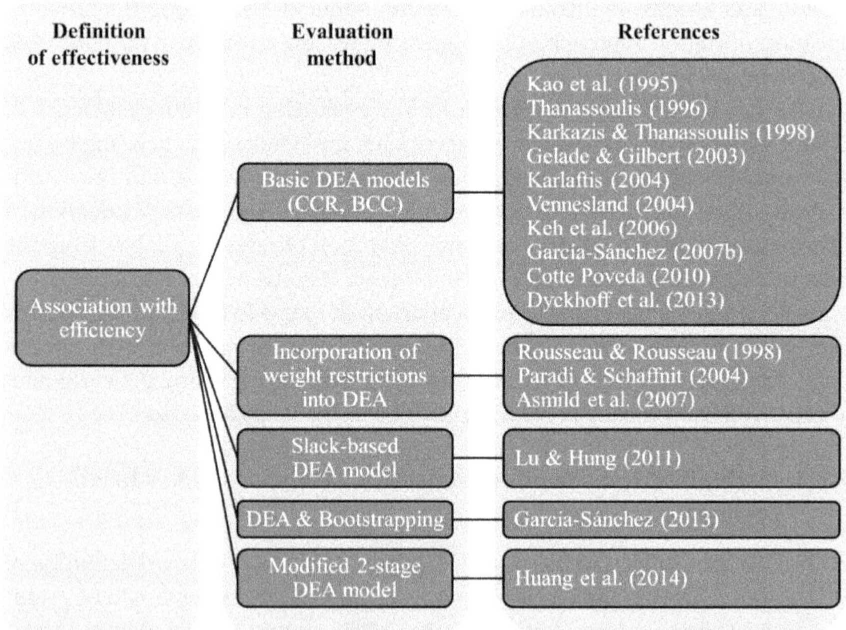

The second category of the reviewed DEA-literature comprises the most contributions, due to the fact that the definition of effectiveness within each concept depends on the context of the corresponding study. The left column in Figure 3.3 provides an overview of different effectiveness concepts, which are referred to in the associated literature that is presented in the last column. Obviously, several new approaches have evolved to evaluate the respective kind of effectiveness. The wide range of the approaches confirms the difficulty of establishing a comprehensive way for the evaluation of effectiveness.

Fig. 3.3: Second category of the reviewed DEA-literature

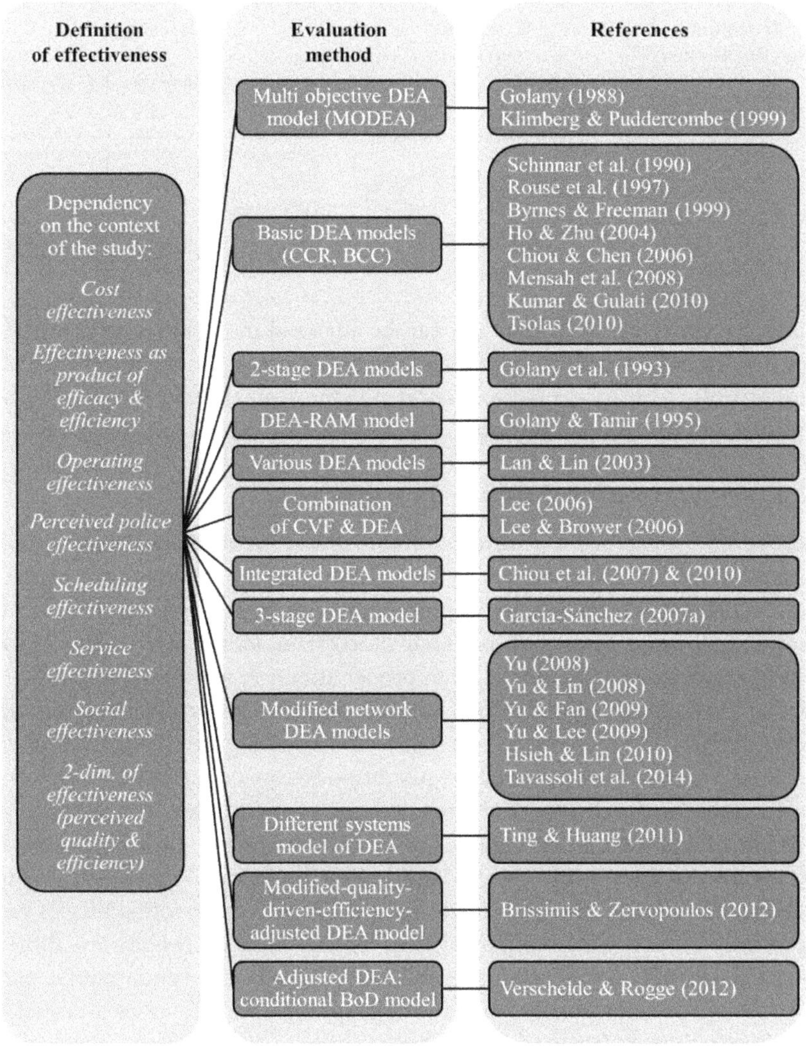

The third and last category entails all references that describe effectiveness as the fulfillment of fundamental goals. There are two approaches that implicitly address this idea: the approaches that integrate DEA and Goal Programming and those eliminating inputs from DEA literature. An overview of the corresponding references is provided in Figure 3.4.

Fig. 3.4: Third category of the reviewed DEA-literature

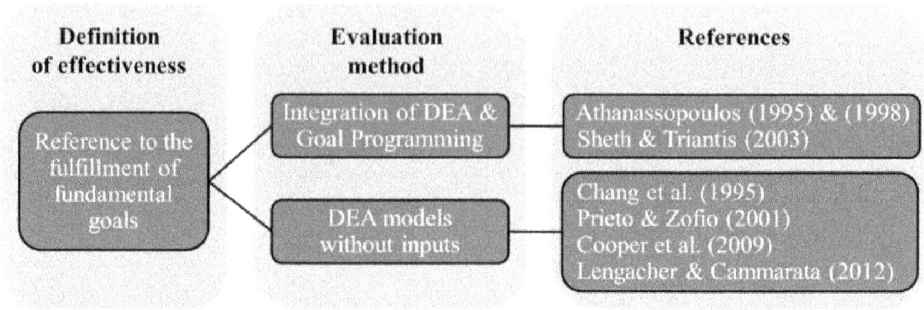

Before the reviewed DEA-literature can be analyzed in detail – especially with regard to the definition of effectiveness that underlies this dissertation – relevant requirements of business practice concerning the evaluation of effectiveness are introduced in the next subsection.

3.2.4 Requirements of business practice

There is often a gap between research findings and their utilization in business practice. The lack of researchers' access to and the experimentation with real-world applications impede the recognition of corresponding research efforts by practitioners. Thus, to prevent research efforts from falling into disuse, practitioners' requirements should be given proper attention within the scientific research.[216] Therefore, the demands of business practice concerning the evaluation of effectiveness, based on the experience from research project with a European pharmacy chain – as a representative practitioner – are taken into consideration as a basis for the development of a relevant and useful model.

Practitioners, such as managers, are forced to regularly evaluate the actual performance of their companies and branches with regard to previously defined strategies in order to maintain the long-term growth and profitability of each organization and organizational subunit. Effectiveness represents one dimension of performance referring thereby exclusively to the fundamental goals pursued by the management and to the respective targets to be attained by the organization. Relating to this understanding of effectiveness, an adequate approach for the evaluation of relative effectiveness should meet the following requirements:

216 Cf. Lewin & Minton (1986), pp. 525–526.

- *Incorporation of predetermined targets as benchmarks*
  In business practice, many companies already use performance management systems, such as balanced scorecard (BSC)[217], that operationalize the fundamental goals by determining appropriate goal indicators and setting targets for each indicator. These targets represent benchmarks for the evaluated companies and/or branches.
- *Consideration of the goal achievement beyond the benchmarks*
  The management usually determines feasible benchmarks, but it also encourages the maximization of goal achievement, i.e. if the organization and/or organizational subunit is/are able to exceed all target levels, this overachievement of the targets has to be taken into consideration and rewarded.
- *Use of a management-oriented approach*
  To ensure the use of the approach – for evaluating the organization's and/or organizational subunits' effectiveness as a complementary measure to its/their efficiency – in business practice, the managers' workload should not be increased and the results should be easy to interpret.

Based on the theoretical recommendations mentioned in Subsection 3.2.2, DEA is considered as a useful approach for evaluating effectiveness, leading to a brief review of current contributions in this area (see Subsection 3.2.3). The capability of these contributions to meet all of the depicted requirements of business practice is critically discussed in the next subsection.

3.2.5 Critical discussion of reviewed DEA-literature concerning the requirements of business practice

The first category of the reviewed DEA-literature associates effectiveness with efficiency (see Figure 3.2). Such a definition is contradictory to the understanding of effectiveness supported by business practice. Therefore, the approaches encompassed by the first category are not suitable for complying with the requirements of business practice.

As can be seen in Figure 3.3, the contributions within the second category are characterized by the diversity of context-dependent definitions of effectiveness leading to a wide range of different models. The requirements of business practice, however, are independent from a specific context referring to effectiveness as the attainment of fundamental goals and respective targets. Consequently, the

217 For more information about the BSC concept, see Kaplan & Norton (1992, 1996).

concepts in this category are inappropriate to meet the requirements mentioned in the previous subsection. Therefore, they are not considered further.

In the last category of the reviewed DEA-literature, the concept of effectiveness is consistent with the one provided by business practice and with that of the underlying dissertation – effectiveness is exclusively referred to the fulfillment of fundamental goals pursued by an organization, representing one of the dimensions of purpose-rational behavior. The contributions that implicitly address this idea are classified into two types: those integrating DEA and Goal Programming and those eliminating inputs from DEA evaluation (see Figure 3.4). These two types of approaches are reviewed in the following with respect to the requirements of business practice previously mentioned.

Approaches combining DEA and Goal Programming incorporate multiple organizational goals and resource allocation strategies in an efficiency-based methodology.[218] The approaches evaluate the performance of a multi-level organization displayed by global levels of inputs/outputs, which are produced by individual DMUs. Each DMU is expected to maximize its contribution to the global targets, which are known a priori by minimizing the negative (underachievement) and positive (overachievement) deviations from the goal targets. The extent to which the organization achieves the global targets is considered as a measure of its effectiveness.[219] However, these approaches involve some limitations, especially when a management-oriented approach is required. From this point of view, the main weakness is the complexity of the approaches, for example, due to the inclusion of enhanced operations research concepts (e.g. Fuzzy Set Theory)[220] or the need for solving a sequence of optimization-based problems to find feasible global targets[221].

Approaches eliminating inputs from DEA evaluation modify traditional DEA models by including only a nominal input constraint. They concentrate on evaluating the extent to which the outputs are achieved; these outputs represent the goals of an organization. For example, Chang et al. (1995) extended the CCR model to evaluate the relative effectiveness of an organization under multiple goals

218 For a detail description of the approaches combining DEA and Goal Programming for evaluation of effectiveness, see Athanassopoulos (1995, 1998) and Sheth & Triantis (2003).

219 Cf. Athanassopoulos (1995), p. 538, and (1998), p. 175–176; Sheth & Triantis (2003), p. 36.

220 By using Fuzzy Set Theory in context of DEA, Sheth & Triantis (2003) developed a DEA approach for evaluating the efficiency and effectiveness in fuzzy environments.

221 Cf. Athanassopoulos (1995), pp. 540–541.

wherein each goal represents a familiar performance ratio. They also combined their model with the Malmquist Productivity Index to measure the effectiveness change. This model provides a management-oriented approach that supports the decision makers with useful information for future goal planning and findings about the improvement potential in key performance ratios. However, this approach does not incorporate predetermined target values of the goals into the analysis and is, therefore, not suitable to measure a goal achievement beyond these targets.

Prieto & Zofío (2001) proposed an approach based on a prior knowledge of goal levels that have to be attained by the DMUs. These goals are interchangeable terms, e.g. if some DMUs exceed the desirable goal levels, they are truncated to define desirable facets of the effectiveness frontier and vice versa. Thus, the reference frontier consists not only of linear combinations of the predetermined, but also of the observed values. The proposed additive range adjusted measure is easy to elaborate if the relevant data is constantly updated. But changes may require a redefinition of the models, meaning that experts familiar with the approach and the management have to collaborate during the complete evaluation process.

Instead of incorporating predetermined goal levels, Cooper et al. (2009) proposed an approach that involves the prior knowledge of experts about the relative importance of the outputs as performance indicators by means of weight restrictions of AR type. The weights represent the decrease or increase rate in effectiveness scores of the corresponding outputs. Using the two-step procedure proposed in Cooper et al. (2007a), they ensured that no output is completely ignored, avoiding the problem of zero-value weights. The usefulness of the approach has been appreciated by basketball experts. However, the incorporation of expert opinions in terms of restrictions on the weights and the subsequent analysis of the results may be difficult to manage without deeper knowledge of the respective methodology.

Lengacher & Cammarata (2012) used the approach provided by Chang et al. (1995) as post-processing effectiveness model to evaluate the relative effectiveness of the set of most-efficient project portfolios. Each portfolio is represented by the sum of cumulative inputs and outputs of the projects contained in this portfolio. Only those portfolios are entering in the post-processing effectiveness model whose cumulative outputs meet or exceed stated goals, whereby the extent to which the outputs exceed the goals is out of consideration. The data provided by the model can be used to create a trade-off curve of efficiency and effectiveness which decision makers can use to identify the win-win bands along the investment continuum. Since it is assumed that the decision makers can influence the allocation of the resources, the approach is most suitable for hierarchical organization in which either the units to some extent have the ability to affect the allocation of the resources or the central unit controls the resource allocation across the units.

A summarizing overview over the contributions within the third category of the reviewed DEA-literature, referring to effectiveness as the fulfillment of fundamental goals, and their compliance with the requirements of business practice is presented in Table 3.1. It is conspicuous that none of the contributions consider goal achievement beyond the benchmarks. This is because in current approaches it is sufficient to meet the goal benchmark. Consequently, a concept capable of evaluating the overachievement of goals would be a pioneer in this field. Furthermore, it is obvious that most of the current approaches are not management-oriented due to their methodological complexity. This confirms the inadequate use of them by practitioners. Better alignment of scientific approaches with the needs of practitioners is, therefore, essential to prevent research efforts from falling into disuse.

Tab. 3.1: Overview over the contributions and their compliance with the requirements of business practice

Contributions referring to effectiveness as the fulfillment of fundamental goals	**Requirements of business practices**		
	Predetermined targets as benchmarks	**Goal achievement beyond the benchmarks**	**Management-oriented approach**
Athanassopoulos (1995)	x	–	–
Chang et al. (1995)	–	–	x
Athanassopoulos (1998)	x	–	–
Prieto & Zofio (2001)	x	–	–
Sheth & Triantis (2003)	x	–	–
Cooper et al. (2009)	–	–	–
Lengacher & Cammarata (2012)	–	–	x

Finally, the detailed critical discussion of the reviewed DEA-literature reveal no approach which meets all practical requirements mentioned in Subsection 3.2.4. Therefore, a new DEA-based approach, which focuses on all of these requirements, is introduced in the following section.

3.3 A DEA approach for evaluating relative effectiveness under predetermined targets

3.3.1 Framework

Two aspects are relevant here for a DEA approach to evaluate relative effectiveness corresponding to the theoretical and practical requirements. First, it is assumed that within the respective organization, there are performance management systems that already operationalize the designated goals by determining goal indicators and setting targets as aspiration levels, allowing to quantifying the achievement of the goals. Therefore, it seems appropriate to use these targets as benchmarks representing a target vector. Second, the goals, goals indicators, and targets are typically "the more, the better"[222] criteria or could at least be transformed into such criteria. This induces the use of an output-oriented approach that excludes the inputs,[223] since only the degree of goal attainment is relevant for an effectiveness analysis, not the extent of resources used.[224]

Two further conditions are relevant in the context of requirements on the part of business practice. On the one hand, it is assumed that the central management of the organization welcomes the achievement of goals beyond the predetermined targets. Characterizing a store of the organization as effective if it reaches all targets, a concept is needed to evaluate the overachievement of goals. Accordingly, a new measure – the *super-effectiveness* score – will be introduced. On the other hand, a management-oriented DEA approach to evaluate the stores' effectiveness is appreciated, which does not increase managers' workload and whose results can be easily interpreted. With reference to this requirement, only the available targets and database will be used to determine the effectiveness of each store, incorporating well-known features of performance management systems.

The resulting two-step procedure to evaluate relative effectiveness is presented in the next subsection. It classifies stores of an organization into subsets according to their ability to meet the targets and it provides a complete ranking within each subset.

3.3.2 Two-step procedure

A common business scenario is considered where an organization is in control of π stores or branches – represented here by the DMUs – with semi-positive input vectors $X^\rho = (x_1^\rho,\dots,x_m^\rho)$ and output vectors $Y^\rho = (y_1^\rho,\dots, y_n^\rho)$, $\rho = 1,\dots,\pi$ for which

---

222 André et al. (2010), p. 374.

223 Cf. Liu et al. (2011), p. 476.

224 Cf. Prieto & Zofío (2001), p. 44.

the central management has determined explicit performance targets by means of the output vector $I = (I_1,\ldots,I_n)$. These targets concretize the overall management goals in terms of expected results to be achieved by the DMUs, providing the basis to evaluate their effectiveness. For this purpose, only the level of goal attainment is relevant, not the extent of resources used, meaning that only the outputs of the DMUs are taken into consideration, not their inputs.

Each DMU manager has to ensure that corresponding DMU is operating in accordance with the specified target values. The actual performance is supposed to be visible by means of a traffic light reporting system. Such an instrument is widely accepted and commonly used in practice: green indicates that a target has been achieved, yellow denotes a performance close to the target, and red represents a performance requiring priority attention due to a significant underachievement of the target.[225] The first step of the procedure refers to this idea of traffic lights in order to classify the DMUs into subsets.

3.3.2.1 *Step 1: Classification as green, yellow, and red DMUs*

$I = (I_1,\ldots,I_n)$ describes the target (output) vector as benchmark defined by the central management. The output vector of $DMU^o = (y_1^o,\ldots,y_n^o)$ is considered where y_i^o represents the value of output i achieved by DMU^o. If this DMU

- fails to meet all goals, i.e.

  $$y_i^o < I_i \quad i = 1,\ldots,n \tag{3.1}$$

  it will be designated as red to highlight that the DMU^o is *ineffective*;
- meets only some goals and others not, i.e. there exists indices $i,t = 1,\ldots,n$ satisfying $i \neq t$, such that

  $$y_i^o \geq I_i \quad \text{and} \quad y_t^o < I_t \tag{3.2}$$

  this will be indicated by the color yellow to express that the DMU^o is *partially effective*;
- meets all goals simultaneously, i.e.

  $$y_i^o \geq I_i \quad i = 1,\ldots,n \tag{3.3}$$

  it will be highlighted by the color green to signal that the DMU^o is *effective*.

225 Cf. Bititci et al. (2000), p. 700; Bourne et al. (2004), p. 381; Schmidt et al. (2006), pp. 128–129.

Since the same target vector is specified for all DMUs under consideration, these restrictions will be identical for all π DMUs. As Figure 3.5 illustrates, this traffic light reporting system subdivides the DMUs into three groups with similar characteristics (ineffective, partially effective, and effective DMUs), but without any further statements about the level of their (in)effectiveness.

Fig. 3.5: Classification of DMUs

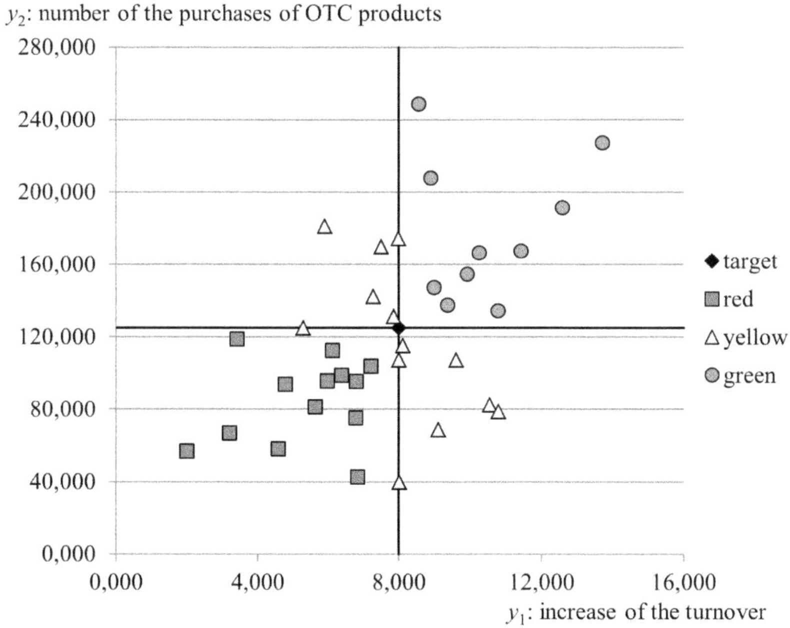

In order to compare the DMUs within the groups – e. g. for rewarding or penalizing them according to their individual goal achievement – it is necessary to determine their degree of relative effectiveness. This is described in the next step of the procedure.

3.3.2.2 Step 2: Determination of relative effectiveness scores

The relative effectiveness scores will provide a complete ranking within the groups of ineffective, partially effective, and effective DMUs. A specific calculation method is provided for each of these groups, taking into account that they differ with regard to their relation to the predetermined targets.

Calculation method for red DMUs

Red (= ineffective) DMUs are addressed first because their effectiveness scores can be calculated on the basis of a conventional DEA approach where only outputs need to be evaluated. Lovell & Pastor (1999) considered the basic DEA models (CCR and BCC models) with a single constant input and compare them to CCR and BCC models without inputs. They prove that an output-oriented CCR (CCR-O) model with a single constant input coincides with the corresponding BCC model. Analogously, an output-oriented BCC (BCC-O) model with a single constant input is equivalent to an output-oriented BCC model without inputs. Thus, a BCC-O model without explicit inputs is applied here:[226]

$$\begin{aligned} & \max\ \eta_*^o \\ \text{s.t.}\quad & \eta_*^o y_i^o \leq \sum_{\rho=1}^{\alpha+1} \lambda^\rho y_i^\rho \qquad i = 1,\ldots, n \\ & \sum_{\rho=1}^{\alpha+1} \lambda^\rho = 1 \\ & \lambda^\rho \geq 0 \qquad \rho = 1,\ldots, \alpha+1 \end{aligned} \tag{3.4}$$

Using the target vector I as benchmark, a set of α red DMUs will obtain scores $\eta_*^0, \rho = 1,\ldots, \alpha$ within the range of (0,1). The DEA results reveal the deficits in the current profiles of the DMUs and provide directions to become (at least partially) effective.

Calculation method for yellow DMUs

Yellow DMUs achieve some of the targets and therefore are characterized by an effectiveness score of 1, which differentiates them from the red DMUs. However, they are just partially effective. In order to quantify the different level of improvement potential of a set of γ yellow DMUs the absolute value of the sum of percentage negative deviations from the predetermined targets is considered as

$$dev^\rho = \left| \sum_{i=1}^{n} \frac{y_i^\rho - I_i}{I_i} \right|, \qquad \rho = 1,\ldots,\gamma \tag{3.5}$$

where dev^ρ is computed only for $y_i^\rho < I_i$. The higher the deviation score of a DMU, the lower is its position in the ranking.

226 Cf. Lovell & Pastor (1999), p. 49.

Calculation method for green DMUs

To discriminate between the green (= effective) DMUs, their overachievement of the given targets can be evaluated with regard to a target vector. Here, a super-effectiveness score is introduced as a specification of this idea. The distance of each green DMU from the effectiveness frontier build by I is evaluated by using the target vector I as a reference point. Due to the fact that the green DMUs are above the effectiveness frontier, they are upwards unlimited, meaning that an output-oriented DEA model is inappropriate for the evaluation of their super-effectiveness scores. Therefore, an input-oriented approach is used in which inputs are exchanged for outputs within the model formulation. The respective modified DEA model for evaluating super-effectiveness reads:

$$\begin{aligned} &\min\ \theta_*^o \\ \text{s.t.}\quad &\theta_*^o y_i^o \geq \sum_{\rho=1}^{\delta+1} \lambda^\rho y_i^\rho \qquad i = 1,\ldots,n \\ &\sum_{\rho=1}^{\delta+1} \lambda^\rho = 1 \\ &\lambda^\rho \geq 0 \qquad \rho = 1,\ldots\ \delta+1 \end{aligned} \tag{3.6}$$

The data set includes the target vector I and δ green DMUs. The reciprocal of the optimal value

$$\psi^o = \frac{1}{\theta_*^o} \tag{3.7}$$

represents the super-effectiveness score of the green DMU^o, ranging from 1 to ∞. The greater the distance of DMU^o from the effectiveness frontier, the higher the super-effectiveness score is. More precisely, the value ψ^o-1 indicates the minimum amount by which DMU^o outperforms all predetermined targets.

3.3.3 Illustrative example

In order to illustrate the proposed procedure, an example based on annual data obtained from a pharmacy chain is presented. The collaborating organization is located in a European country in which the prevailing legal norms allow the formation of pharmacy chains. Thus, the pharmacy chain was able to strengthen its position in the market enormously in the last ten years. Its acquisition of single pharmacies leads to a portfolio of over 100 pharmacy stores that have to be managed effectively and efficiently in order to maintain the long-term growth trend and profitability.[227]

The individual pharmacy stores operate in different environments in terms of location, the respective clientele, and fields of activity. An effective management of the whole portfolio requires the consideration of the specific environmental characteristics. Therefore, the pharmacy stores are divided into three homogeneous groups according to their location:

- *Community pharmacies*, located in suburban areas, usually offer their range of services to a constant customer base. These pharmacies benefit from the advantage that in such areas the number of competitors is limited. By establishing closer relationships with the customers, the pharmacies focus on the achieving customers' satisfaction in order to become their first point of contact. This strategy enables the pharmacies to adapt their range of services to the specific needs of their clientele and to simultaneously reduce the storage costs of products that are out of demand for their customers.
- *Hospital pharmacies* are located in hospitals or clinics. The collaboration with the physicians and other health care professionals in various medical areas demands of the hospital pharmacists to gain more education and training in the complex clinical medications. Moreover, they have to stock a wider range of medicines. The customers of the hospital pharmacies are mostly the hospital patients. Consequently, the customer loyalty is dependent on the patients' state of health and their sojourn time in the hospital or clinic. On the one hand, this means that the pharmacies are not able to establish a long-time relationship with their customers in order to strengthen the customer loyalty. On the other hand, this also means that the pharmacies are operating in an almost non-competitive environment.
- *Shopping center pharmacies* are located in cities within shopping centers or malls. These pharmacies are characterized by an intense competitive environment. They have to compete with other pharmacy stores in the immediate surroundings as wells as with the growing number of non-pharmacy retailers, such as drug stores, perfumeries, or grocery stores. Additionally, it is all but impossible to build up a group of regular customers in such highly frequented locations. To stay competitive, the pharmacies either have to eliminate the potential sources of loss of possible customers, e.g. long waiting times, or have to distinguish themselves from the competitors by a specialization in a supplemental service like homeopathy or nutrition support.

227 In the following, the terms "pharmacy" and "pharmacy store" are used as synonyms.

Each of these groups is supervised directly by the central management, that means the pharmacy chain adopts a very flat hierarchical structure. It consists of only two management levels: the central management at the top – responsible for the economic performance of the pharmacy chain – and the store managers on the first-level – responsible for the operational activity of their pharmacy stores. According to the conditions in the operating environment of each group, central management determines appropriate managerial strategies and goals – which are operationalized by measurable performance indicators – and set targets for each goal to be pursued by the pharmacy stores. As a result, different goal systems are developed, one for each group. These goal systems usually overlap each other due to the fact that the majority of the goals is valid for the whole portfolio of pharmacy stores. Representative examples are the financial goals such as "maximization of gross profit" or "ensuring long-term profitability".

This example focuses on the evaluation of the group with the strongest competitive environment: shopping center pharmacies. Only those stores are analyzed for which data about their outpatient care has been available. The resulting sample consists of 36 pharmacies with similar responsibilities.

Based on the decision guidelines provided by Cameron & Whetten (1983b), the scope of the assessment is described in order to explain what kind of effectiveness is being evaluated. The seven decision guidelines and corresponding circumscriptions are summarized in Table 3.2.

For the purpose of the assessment, the following three customer-oriented goals are selected:

- *Exploitation of customer potential*
  goal indicator: absolute increase of the turnover (in Tsd. €);
- *Strengthening of customer retention*
  goal indicator: number of the purchases of over-the-counter products;
- *Increase of popularity*
  goal indicator: number of the responses to an annual competition in each store's catchment area.

The corresponding data set of the 36 pharmacy stores together with the associated target vector I is shown in Table 3.3.[228] Subsequently, Figure 3.6 illustrates the current levels of goal achievements of all 36 shopping center pharmacy stores in comparison with the corresponding target. Green DMUs, A1 to A9, meets or

228 Each pharmacy store represents one DMU. For simplicity, the DMUs are labeled as A1 to A36.

exceeds the targets, yellow DMUs, A10 to A24, meets some of the targets but not all three, and red DMUs, A25 to A36 meet none of the targets.

Tab. 3.2: Decision guidelines and the descriptions of the assessment

Decision guideline	Description
1. From whose perspective is effectiveness being judged?	The strategic management level of the pharmacy chain, i.e. the central management, constitutes the relevant perspective in this example. As the major decision maker, it has the most influence on the pharmacy stores' direction and performance.
2. On what domains of activity is the judgment focused?	The major area of activity and identity for each of the pharmacy stores is the provision of pharmaceutical care to the customers. The quality of the delivery of customer services is a comparable domain across pharmacy stores operating in similar environments.
3. What level of analysis is being used?	The subunit level of analysis is the focus. This level allows comparative judgments across the subunits, i.e. pharmacy stores. However, the pharmacy stores are divided into three homogeneous groups, meaning that each group should be evaluated independently from the others. Here, the shopping center pharmacy stores are exemplarily chosen for the assessment.
4. What is the purpose of judging effectiveness?	The purpose of the assessment is to identify improvement potential concerning the ability of the pharmacy stores in providing pharmaceutical care to existing customers and simultaneously attracting as well as retaining new customers.
5. What time frame is being employed?	Short-term (annual) effectiveness in delivering high standards of customer services.
6. What type of data is being used for judgments of effectiveness?	The central management controls the pharmacy stores by determining goals and operationalizing them via goal indicators. These goal indicators are derived from the internal data of organizational records, e.g. financial data from accounting systems, as well as non-financial data obtained from marketing operations.
7. What is the reference against which effectiveness is judged?	The effectiveness is judged against a goal-centered reference, here, a predetermined target vector.

Tab. 3.3: Data set of 36 pharmacy stores and the target vector I

DMU	Goal 1: exploitation of customer potential	Goal 2: strengthening of customer retention	Goal 3: increase of popularity
	TURN: increase of the turnover	OTC: number of the purchases of over-the-counter products	RES: number of the responses to an annual competition in the catchment area
I	I_1: 8,000	I_2: 125,000	I_3: 15,000
A1	9,938	154,500	20,000
A2	11,439	167,450	18,490
A3	13,726	227,050	16,930
A4	8,917	207,650	17,130
A5	12,604	191,250	16,140
A6	10,802	134,250	18,000
A7	8,576	248,750	15,650
A8	9,006	147,200	15,500
A9	10,273	166,250	15,160
A10	7,511	170,000	14,180
A11	9,381	137,500	12,550
A12	9,612	107,500	13,860
A13	7,990	174,550	10,920
A14	7,862	131,450	10,380
A15	10,561	82,500	14,840
A16	7,281	142,500	10,110
A17	5,899	181,250	12,230
A18	5,981	95,600	17,860
A19	8,000	107,500	9,460
A20	5,295	125,000	12,390
A21	8,120	115,400	6,940
A22	10,799	78,900	9,870
A23	9,113	68,850	5,800
A24	8,014	40,000	7,830
A25	6,388	98,750	14,670
A26	6,791	75,000	14,510
A27	6,849	42,500	14,350
A28	3,424	118,750	11,580
A29	5,640	81,250	14,020
A30	3,211	66,750	13,730
A31	7,223	103,750	10,920
A32	6,130	112,500	8,480
A33	2,010	56,700	12,840
A34	6,797	95,250	2,060
A35	4,806	93,750	11,250
A36	4,598	58,100	10,050

Fig. 3.6: Current levels of goal achievements of the 36 pharmacy stores

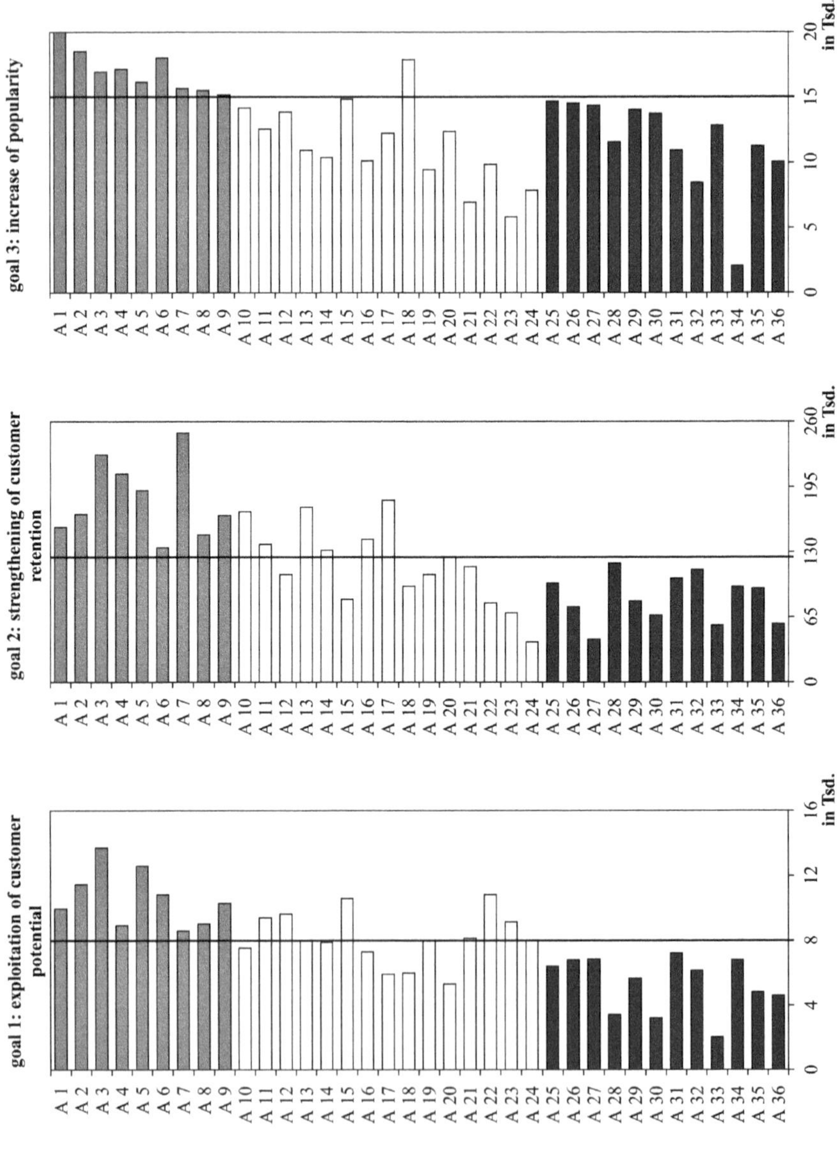

Tab. 3.4: Results of the effectiveness evaluation procedure

Classification	DMU	Ranking	Effectiveness score
Green: effective	A1	1	1.24
	A2	2	1.23
	A3	3	1.13
	A4	4	1.11
	A5	5	1.08
	A6	6	1.07
	A7	7	1.04
	A8	8	1.03
	A9	9	1.01
Target	*I*	1	1
Yellow: partially effective	A10	1	1 (11.6%)
	A11	2	1 (16.3%)
	A12	3	1 (21.6%)
	A13	4	1 (27.3%)
	A14	5	1 (32.5%)
	A15	6	1 (35.1%)
	A16	7	1 (41.6%)
	A17	8	1 (44.7%)
	A18	9	1 (48.8%)
	A19	10	1 (50.9%)
	A20	11	1 (51.2%)
	A21	12	1 (61.4%)
	A22	13	1 (71.1%)
	A23	14	1 (106.3%)
	A24	15	1 (115.8%)
Red: ineffective	A25	1	0.98
	A26	2	0.97
	A27	3	0.96
	A28	4	0.95
	A29	5	0.93
	A30	6	0.92
	A31	7	0.90
	A32	8	0.90
	A33	9	0.86
	A34	10	0.85
	A35	11	0.75
	A36	12	0.67

Table 3.4 presents the main numerical results of the proposed procedure, i.e. the classification related to the traffic light reporting system, the effectiveness scores and the resulting ranking for each category:

- *Green DMUs:*
  A1 to A9 are characterized as super-effective because they overachieve all predetermined targets. For example, DMU A1 obtains the highest super-effectiveness score of 1.24, meaning that it outperforms all targets by at least 24%; a closer look at the data reveals that it outperformed I_1 by 24%, I_2 by 24% and I_3 by 33%.
- *Target vector:*
  As its own category, the target vector I represents the benchmark that concretizes the overall management goals in terms of expected results, which should be achieved by the DMUs. Therefore, its effectiveness score is 1.
- *Yellow DMUs:*
  A10 to A24 are partially effective since they meet not all, but at least one of the targets. In accordance with the idea of setting a DMU in the best possible light, their effectiveness score is 1. In order to differentiate within this group of DMUs, a ranking is provided based on the absolute value of the sum of percentage negative deviations from the predetermined targets. A10 can be found on the first position due to its lowest deviation score of 11.6%
- *Red DMUs:*
  A25 to A36 are characterized as ineffective as they do not reach any target. For instance, A36 obtains an (in)effectiveness score of 0.67, expressing that it achieves not more than 67% of each target set by the central management.

3.4 Managerial implications

Characterizing effectiveness as an independent performance dimension, which refers exclusively to the fundamental goals of economic activities, allows to conduct a broader performance assessment. Regardless of the aspect of efficiency, the fulfillment of fundamental goals is of vital importance for long-term business success of an organization. The results of the procedure proposed here for evaluating effectiveness support managerial decision making in several ways. They are comprehensible and create transparency regarding the effectiveness of the DMUs. The performance of a certain DMU compared to the others and to the target vector I can be easily appraised. The traffic light classification provides a quick but well-founded insight into the overall performance structure, indicating whether the target vector is appropriate or too (un)ambitious. Thus, the results also have an impact on performance control. Targets that are not achieved at all or over-fulfilled by almost all DMUs should be reconsidered together with the corresponding goals.

The complete ranking within each subset – effective (green), partially effective (yellow), and ineffective (red) DMUs – can be used as a basis for an incentive system. Rankings based on the results of traditional DEA models may be considered as inappropriate due to the differences in the underlying reference sets. In detail, the methodology of DEA determines a piecewise linear surface as frontier consisting of best performing DMUs that represent benchmarks for the underperforming DMUs. Since DEA aims to assign each DMU its best possible score, the corresponding benchmarks are usually derived from the frontier facets closest to the respective DMU. As a consequence, the benchmarks may differ from DMU to DMU. The same is true for the flexibility in the choice of weights for performance criteria, impeding the use of them as a basis for a ranking. Thus, it is generally not advisable to use the traditional DEA results for ranking.[229] However, the proposed procedure overcomes the mentioned problem by using the predetermined target vector *I* as a consistent benchmark for all DMUs within the sample. At the same time, the different models to calculate the effectiveness scores of red, yellow, and green DMUs induce plausible incentives.

Particularly, an ineffective DMU may concentrate its efforts on one or a few goals, which are already fulfilled better than the others, in order to quickly reach a higher ranking position (or even become partially effective). This strategy no longer works for a partially effective DMU. In order to reach a higher ranking position, such a DMU has to focus on those goals for which it does not achieve the targets. In other words, the overachievement of targets does not compensate the underachievement of other targets. Finally, a super-effective DMU can increase its effectiveness score only if it improves those goals with the relatively lowest overachievement of the target values. In this respect, the proposed procedure implements the suggestion of Cameron (1986) that an organization should focus on balance concerning its goals and corresponding criteria.

3.5 Conclusion and discussion

This chapter has addressed the managerial and conceptual problem of defining and evaluating the relative effectiveness of DMUs. The conceptual disorder in the pertinent literature has been described by considering the evolution of organizational effectiveness and the major theoretical approaches. Based on the resulted findings that goals are adequate standards for the definition and evaluation of effectiveness and following Weber's (1978) treatise on purpose-rational behavior, effectiveness has been characterized as an independent performance dimension, which refers exclusively to the original goals of economic activities.

229 Cf. Sinuany-Stern & Friedman (1998), p. 472; Adler et al. (2002), p. 255.

Concerning the evaluation of effectiveness, substantial problems that are still inherent in evaluating effectiveness have been discussed. Recommendations provided in the literature regarding these problems have been presented and complemented by relevant requirements of business practice. Since DEA is recognized as a potentially useful approach for the evaluation of effectiveness, the existing contributions regarding the evaluation of effectiveness by means of DEA have been reviewed. The review has not revealed a suitable approach that meets all described practical requirements. Therefore, a new DEA approach for the evaluation of effectiveness in the presence of predetermined targets, which are to be achieved simultaneously, has been developed. This approach differs from the other DEA approaches dealing with the evaluation of effectiveness in three aspects:

- As a management-oriented approach, it is based on the well-known traffic light reporting system and incorporates information about predetermined targets. Such information is usually available in the context of common performance management systems like the BSC.
- Based on the target vector *I* as a benchmark for all DMUs, it classifies them into subsets according to their ability to meet the targets and provides a complete ranking within each subset. Thereby, the concept of super-effectiveness is introduced to measure the effectiveness of those DMUs which over-fulfill all goals.
- Both decision facilitating and decision-influencing purposes are taken into account. Concerning the last-mentioned aspect, the ranking models induce certain strategies of the DMUs' management in order to achieve a higher ranking position, leading towards a balanced fulfillment of the targets.

The introduced approach is not restricted to the specific application of pharmacy stores, as shown in this chapter's example, or to the presence of a BSC but can be applied to a broad field of cases where the same targets are set for all DMUs under consideration.

From a managerial point of view, the new approach is aimed to meet the requirements on model design provided by Little (2004). Its strengths and limitations referring to these requirements are presented in Table 3.5. In detail, the approach is simple to understand due to the comprehensible evaluation process resulting in a single aggregate measure of relative effectiveness that does not require a priori specification of a weighting scheme. The approach can be characterized as robust as long as the assumption of one predetermined target vector is valid. In case of more than one predetermined target vector, a modification of the procedure is necessary, e.g. based on the layer model of Alirezaee & Afsharian (2007).

Tab. 3.5: Strengths and limitations of the presented DEA approach for evaluating the relative effectiveness of DMUs under predetermined targets

Managerial requirements on model design[230]	Strengths	Limitations
Simple	Single aggregate measure of relative effectiveness for each DMU No a priori specification of weights for goals and/or effectiveness criteria Comprehensible background of the evaluation process	
Robust	Replicable measure of relative effectiveness based on the comparison with the target vector	Not suitable in case of more than one target vector
Easy to control	Incorporation of information usually available in the context of common performance management systems Classification in comparable groups (effective, partially effective, ineffective)	Existence of predetermined goals and corresponding targets is required
Adaptive	Different units of measurement are allowed	Potential danger of mixing short with long term measures as well as indices with volume measures
Complete on important issues	Simultaneous consideration of multiple goals and/or effectiveness criteria	Subjectivity concerning the selection of goals and/or effectiveness criteria
Easy to communicate with	Quick computation of effectiveness scores and provision of rankings	

The most beneficial strengths of the approach lie in the incorporation of information about goals and targets usually available within the organizations and the classification of DMUs in comparable groups according to the modified concept

230 Cf. Little (2004), pp. 1843–1844.

of traffic lights in the case of effectiveness evaluation. Although, this property enables the approach to be easy to control, it leads to the dependence of the approach on the existence of such information. Otherwise, the managers' workload would increase in order to identify and determine suitable goals and targets for the evaluation of effectiveness.

The possibility to handle different units of measurement enables the introduced DEA approach to be adaptive to new information. However, the problem of mixing short-term with long-term measures as well as indices with volume measures mentioned by Dyson et al. (2001) still exists. This concerns especially the effects of some short-term goals on the long-term goals, which are difficult to measure and to eliminate. Here, it is extremely necessary that only the fundamental goals are chosen for the evaluation of effectiveness and not the instrumental ones.

The proposed approach is complete on important issues in the sense that it enables to simultaneously consider multiple goals and/or effectiveness criteria as well as targets. Nevertheless, the decision about the incorporation of goals and/or effectiveness criteria into the evaluation is influenced by subjective judgments of the decision makers. Subjectivity is a general problem in the context of performance evaluation that is impossible to eliminate totally but can be reduced through the consultation of different decision makers.

Finally, the computation process of the underlying approach makes it possible to easily get alternative results, e.g. the decision makers can vary the combination of goals and targets to see the resulting effects on the effectiveness scores. Due to this property the approach seems to be intuitive, increasing the willingness of the decision makers to really use this approach in practice.

4 Evaluation of relative balance[231,232]

Performance evaluation, as an essential part of performance management, relies on a careful selection of a set of appropriate and measurable performance criteria, which best capture the performance of the organization or organizational subunits under consideration. Due to the need of parsimony, these set of performance criteria entails only the most important and meaningful measures. Obviously, it is expected that all of the selected criteria are included in the subsequent performance evaluation. The possibility of neglecting some of the performance criteria within the evaluation is therefore considered as contradictory and may be confusing for decision makers.[233]

DEA as a meaningful approach to relative efficiency and effectiveness evaluation of DMUs leads to such a significant shortcoming. It allows zero-value weights that eliminate from the analysis any input and/or output criteria in which the performance of a DMU is weak, with the aim of raising its efficiency score – and effectiveness score respectively – to its maximum level.[234] DMU managers who are responsible for the efficiency as well as effectiveness of the DMUs may appreciate this property of DEA, but higher management might not.[235] In fact, the possibility of neglecting important inputs and/or outputs is a considerable source of pitfalls, not only in performance assessment but also in performance control. While the first aspect refers to a questionable evaluation of efficiency and effectiveness themselves, the second aspect refers to the usually undesirable incentive to concentrate management efforts on a few performance criteria.[236] Both aspects constitute main barriers to the acceptance of DEA in practice.

---

231 The content described in this chapter has been partially published by Ahn, H., Neumann, L., & Vazquez Novoa, N. (2012): Measuring the relative balance of DMUs, in: European Journal of Operational Research 221(2), pp. 417–423. The content contains some minor changes in comparison with the published version.

232 Some parts of the content described in this chapter has been published by Ahn, H., Neumann, L., & Vazquez Novoa, N. (2012): Relative balance as complementary measure to relative efficiency, in R.D. Banker, A. Emrouznejad, A.L.M. Lopes, & M.R. de Almeida (Eds.): Data Envelopment Analysis – Theory and Applications: Proceedings of the 10th International Conference on DEA (pp. 265–271). August 2012, Natal, Brazil.

233 Cf. Bougnol et al. (2012), pp. 698–699.

234 Cf. Dyson & Thanassoulis (1988), p. 564; Allen et al. (1997), pp. 14–15; Thanassoulis et al. (2004), p. 100.

235 Cf. Farzipoor Saen (2010), pp. 2820–2821.

236 Cf. Pedraja-Chaparro et al. (1997), p. 218; Allen & Thanassoulis (2004), p. 363; Estellita Lins et al. (2007), p. 956.

A series of articles addressing the problem of zero-value weights has been published, originating a research area called *value judgment*[237] in DEA. The approaches proposed to avoid unacceptable weighting schemes are varied and encourage the practical application of DEA. However, these approaches are not free from considerable disadvantages. Most approaches require additional subjective information concerning the respective evaluation case, thus increasing managers' workload and the risk of misleading decisions. The aim of this chapter is to introduce a new approach that complements the two common aspects of performance – efficiency and effectiveness – with a measure of *balance*, as an additional dimension of performance. Without the need for further value judgments, the respective balance score evaluates the extent to which a DMU avoids concentration on only some of the crucial performance criteria and goals respectively, providing thereby important insight into a DMU's performance structure.

This chapter is organized as follows. Section 4.1 detects the managerial problems resulting from the flexibility in weight determination which in turn constitutes one of DEA's major strengths. A brief review of value judgment literature reveals no overall satisfactory solution to the limitations stemming from flexible weights. Therefore, the idea of a complementary performance measure – the balance score is presented – that directly refers to the given values of the crucial performance criteria, avoiding thereby further value judgments. Section 4.2 introduces the corresponding DEA approach for the evaluation of relative balance of DMUs and illustrates the approach by continuing the example of shopping center pharmacies presented in Chapter 3. The complementary information provided by the balance scores can have various and substantial implications for management, which are discussed in Section 4.3, leading to an iterative evaluation process of balance and efficiency. Since the approach introduced in Section 4.2 has already been published as Ahn et al. (2012a), subsequent model developments are briefly reviewed in Section 4.4. Finally, Section 4.5 concludes the chapter and discusses the characteristics of the introduced approach with regard to the fulfillment of requirements on model design from the managerial point of view.

237 Allen et al. (1997), p. 14, defined value judgments as "logical constructs, incorporated within an efficiency assessment study, reflecting the decision makers' preferences in the process of assessing efficiency".

4.1 Flexible weights in DEA models

4.1.1 Managerial problems resulting from flexible weights

Since the appearance of the first publications addressing the perceived worth of inputs and outputs in DEA models, the importance of value judgment in DEA has significantly increased. This research has been motivated by different applications of DEA to several real life problems, varying from evaluations of highway maintenance patrols and hospitals, to analyses of the steel industry and bank branches. In such cases, traditional DEA multiplier models were found to allow too much flexibility for the determination of the value system reflected by the weights of the inputs and outputs.[238]

Dyson et al. (2001) systematically described the pitfalls originating from the flexibility of the weights. One of the shortcomings they highlighted is the permission of (nearly) zero-value weights, excluding the respective performance criteria from the efficiency evaluation. Dyson et al. (2001) argued that this is contradictory to the

> [...] agreement on which factors to include in the DEA assessment.[239]

In other words, neglecting those performance criteria that have been previously defined as being relevant can significantly limit the validity of the assessment.[240]

In addition to the matter of performance assessment, a further problem is related to performance control. Zero-value weights may lead to a misallocation of resources, as DMU managers could attempt to raise the unit performance by concentrating their efforts only on a few performance criteria. This dysfunctional incentive will be reinforced if central management rewards DMUs according to their DEA efficiency score. As a result, DMUs will show a tendency to become unbalanced with regard to their performance criteria values, i.e., at least one of the relevant criteria will be neglected in favor of the others. This in turn will decrease the motivation of managers who run balanced DMUs. Their efforts towards achieving an adequate level on all relevant criteria imply the risk of not reaching the highest possible DEA efficiency score. Eventually, strategic conflicts for the organization as a whole may arise.[241]

---

238 Cf. Thanassoulis et al. (2008), p. 322.

239 Dyson et al. (2001), p. 253.

240 Cf. Roll et al. (1991), p. 3; Pedraja-Chaparro et al. (1997), pp. 218–219; Thanassoulis & Allen (1998), p. 586.

241 Cf. Epstein & Henderson (1989), p. 114; Ittner et al. (2003a), p. 754; Sprinkle & Williamson (2007), pp. 430–431.

The literature to date provides various approaches for incorporating value judgments in DEA. In the next subsection the main broad types of theoretical approaches are briefly reviewed.

4.1.2 Value judgment in DEA: a brief review

To address the limitations stemming from the flexibility of weight determination, two types of approaches have been proposed: those imposing restrictions on the weights and those changing the comparative set of DMUs.[242] A recent overview of these approaches, which are gathered under the term of value judgment, has been presented by Thanassoulis et al. (2008).[243]

Approaches involving *weight restrictions* (WRs) have been described by Roll et al. (1991) as

> [...] a step towards switching from the points of view of the individual DMUs only, to that of central management which controls the entire group of the DMUs.[244]

These approaches usually require additional information about constraints and therefore augment the subjectivity of the analysis.[245] Moreover, WRs approaches are criticized because of the difficulties in interpreting radial efficiency measures[246] and are characterized as being too restrictive[247].

There are two main groups of WRs:[248] those restricting the virtual inputs and outputs (virtual weights) and those directly imposing constraints on the weights in the multiplier model. The latter are less complex, but they involve a number of limitations, including the difficulty of comparing different units of measurement and the risk of infeasibility.[249] Additional limitations include sensitivity to model orientation and the need to set individual WRs for each DMU.[250] Approaches imposing restrictions on the virtual weights solve the problem regarding the combination of different units of measurement but are usually computationally

---

242 Cf. Allen & Thanassoulis (2004), pp. 363–364.
243 Cf. Thanassoulis et al. (2008), pp. 321–340.
244 Roll et al. (1991), p. 8.
245 For an exception, see the model proposed by Ramón et al. (2010).
246 Cf. Allen et al. (1997), p. 27.
247 Cf. Allen & Thanassoulis (2004), p. 372.
248 Cf. Thanassoulis et al. (2008), pp. 322–332.
249 Cf. Sarrico & Dyson (2004), p. 18; Estellita Lins et al. (2007), p. 957; Khalili et al. (2010), p. 761.
250 Cf. Allen et al. (1997), p. 19; Pedraja-Chaparro et al. (1997), p. 221.

expensive. In addition, they may still entail sensitivity to the model orientation and the risk of infeasibility.[251]

Approaches *changing the comparative set of DMUs* either modify the input-output data of the DMUs or add unobserved DMUs to the data set.[252] The approaches that directly act on the input-output data allow the computation to be handled by the traditional radial models; however, they may be affected by infeasibility, and the efficient radial targets may not always be technologically possible.[253] Adding unobserved DMUs to the original data set improves the envelopment of inefficient DMUs by preventing them from being projected onto non-efficient facets of the frontier, i.e. assigning a zero-value weight to some inputs and/or outputs. For this reason, some authors consider this approach to be a consequence of facet extension.[254] The main difficulty emerging from the introduction of unobserved DMUs relates to the creation of these DMUs; revealing necessary information about local preferences may be a time-consuming task for management. Furthermore, adding unobserved DMUs also bears the risk of infeasibility and may be ineffective if many DMUs make reference to non-full dimensional efficient facets.[255]

Figure 4.1 provides a structured overview of the two main types of approaches incorporating value judgments and their limitations. The review reveals that the value judgment literature to date provides no overall satisfactory solution to the zero-value weights problem. In addition to the mathematical limitations, managers are expected to possess a deep knowledge of the respective DEA techniques regarding the requirement of additional subjective information about model restrictions and/or parameters. Not only does this increase managers' workload, but it also increases the risk of misleading judgments.

To cope with the problems resulting from zero-value weights without the need for further value judgments, a complementary performance measure that directly refers to the given values of the crucial performance criteria is introduced in the next subsection.

251 Cf. Allen et al. (1997), p. 24; Thanassoulis et al. (2004), pp. 114–115; Tracy & Chen (2005), p. 391; Estellita Lins et al. (2007), p. 957.
252 Cf. Thanassoulis et al. (2008), pp. 333–339.
253 Cf. Hinojosa & Mármol (2011), p. 160.
254 Cf. Portela & Thanassoulis (2006), p. 136; Cook & Seiford (2009), p. 9.
255 Cf. Allen & Thanassoulis (2004), p. 373; Thanassoulis et al. (2004), p. 131.

Fig. 4.1: Overview of the two main types of approaches incorporating value judgments and their limitations

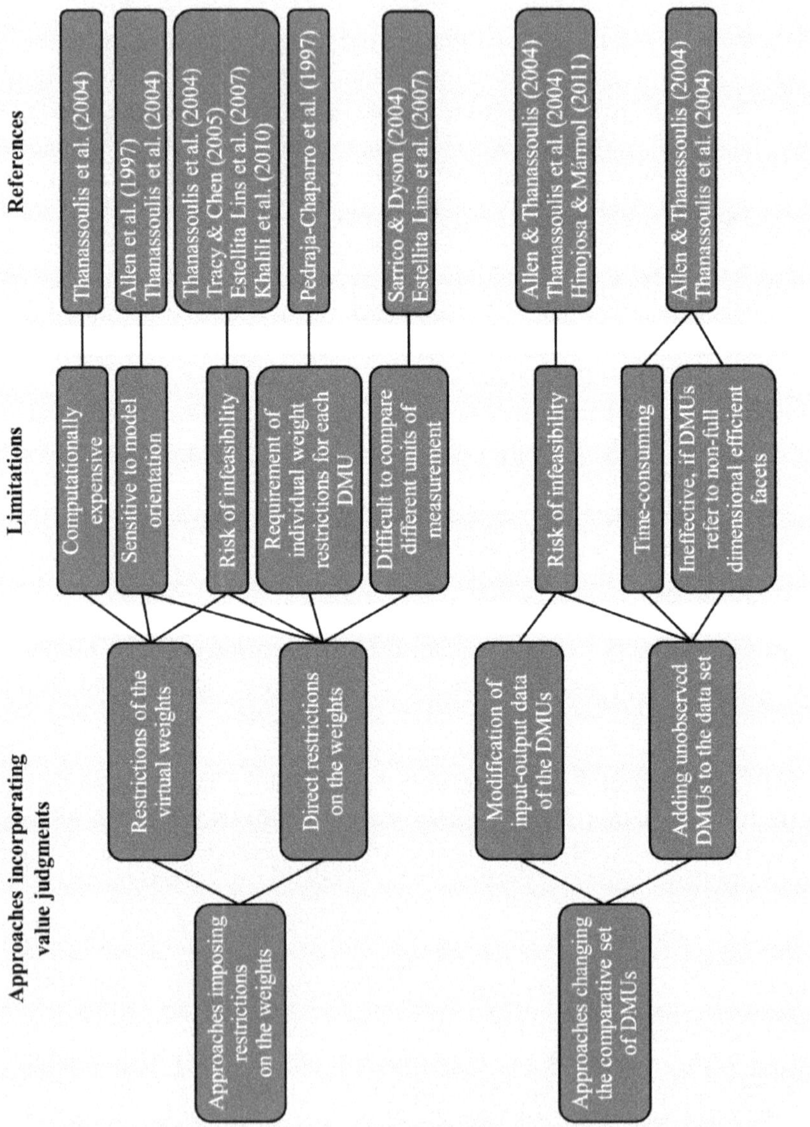

4.1.3 Relative balance as an additional dimension of performance

Weights are inevitable in the context of a performance evaluation, since multiple performance criteria – usually defined in different units of measurement – have to be combined to provide an overall picture of performance.[256] Subjectivity in performance criteria weighting may influence the results of the performance evaluation and prevent thereby their acceptance as fair and equitable. Approaches, like DEA, which allow flexibility in weight determination instead of the incorporation of subjective weights, seem to be more appropriate for an accurate performance evaluation. However, this characteristic bears also a significant shortcoming: it allows the neglect of performance criteria in which the performance of a DMU is weak. This is contradictory to the basic assumption of the performance evaluation in which the performance of a DMU is evaluated by taking into account all of the selected performance criteria. As a consequence, the possibility of neglecting important performance criteria may induce extremism in the criteria achievement of the DMUs, resulting in a questionable evaluation of their performance. To prevent undesirable incentives of concentrating management efforts only on a few performance criteria, the inclusion of all pre-selected performance criteria should be ensured within the evaluation.

One of the recent approaches, which address the problem of excluding criteria from the performance evaluation, has been presented by Dimitrov & Sutton (2010). The authors introduced a scaling factor, which

> [...] determines how much a particular DMU will be penalized for an asymmetric selection of virtual weights.[257]

As a major advantage, their model does not affect feasibility. However, the value of the non-negative scaling factor must be determined for each specific evaluation case. On one hand, this allows for variation in the extent to which a DMU is penalized for asymmetric weight selection. On the other hand, it also forces the management to decide on the appropriate factor value.[258]

An alternative approach for "penalizing" DMUs that only reach their best possible efficiency score through an extreme selection of weights is proposed here. Without the need for further value judgments, the presented approach comple-

256 Cf. Sprinkle & Williamson (2007), p. 430.
257 Dimitrov & Sutton (2010), p. 284.
258 Cf. Dimitrov & Sutton (2010), pp. 284–285.

ments the two common aspects of performance – efficiency and effectiveness – with a measure of balance[259] as an additional dimension of performance.

Balance concerning the goals and corresponding performance criteria of a DMU has already been an important topic in Chapter 3.[260] While inputs are usually characterized as a means to achieve the fundamental goals of an organization, outputs represent measurable indicators for these goals. Therefore, the pattern of the input values is estimated to be of minor relevance, whereas the pattern of the output values is estimated to be of great managerial importance. As it may be the case for a range of organizations and according to the abovementioned basic assumption of the performance evaluation, it is assumed that managers prefer that attention is paid to the achievement of all goals (or outputs respectively), instead of neglecting some goals (or outputs respectively) to raise the level of the others. Without further value judgments, the DEA efficiency score is not able to reflect such preferences concerning the value mix of the outputs as crucial performance criteria. It may happen that different DMUs obtain (nearly) the same efficiency score, even if they substantially differ in their pattern of criteria achievement. Therefore, this aspect of performance is covered by the balance score:

> *Balance:* A DMU will be characterized by a balance score of 100% if all of its crucial performance criteria values are – relative to the respective values of the other DMUs – achieved on a similar level. In contrast, the more the performance criteria values are not achieved on a (relative) similar level, the more the balance score decreases towards 0%.

The balance score is to be understood as a complementary measure to the traditional efficiency and effectiveness scores. It constitutes a source of additional information about a DMU's performance structure that not only permits the evaluation of the comparability of the DMUs under analysis but also questions the role of efficient but unbalanced DMUs. Before such implications for the management can be discussed in detail, the following section introduces – as far as is known – the first DEA approach for the evaluation of relative balance of DMUs.

---

259 The basic idea of balance as an additional performance measure originates from Heinz Ahn (Technische Universität Braunschweig) and Harald Dyckhoff (RWTH Aachen University) and arises from a joint research project "Advanced Data Envelopment Analysis" supported by German Research Foundation.

260 With regard to the theoretical explanation of the need for balance in performance criteria in the context of the evaluation of relative effectiveness, see Subsection 3.1.2. For managerial implications resulting from the incorporation of some kind of balance in effectiveness measurement, see Section 3.4.

4.2 Innovative DEA approach for evaluating the relative balance of DMUs

4.2.1 Framework

Three aspects are relevant here for a new approach to evaluate the relative balance. Firstly, in order to keep the managers' workload to a minimum, the approach should not require any additional information to quantify the balance scores. Thus, the balance score is derived from the given data set of a standard DEA application, i.e. it directly refers to the given input/output values. Secondly, it is assumed that a DMU is generally able to substitute its outputs, one for the other in line with its capacity. This implies that through a reduction of an output, which is produced to a large amount, the production of another output – which actual amount is too small – can be increased. Thirdly, the approach intends that the balance score supplements the well-known aspects of performance – efficiency and effectiveness – with new insights into a DMUs performance structure. Therefore, the approach is independent from the efficiency and effectiveness evaluation by means of DEA.

The resulting two-step procedure to evaluate the relative balance is presented in the next subsection. It classifies the DMUs in balanced and unbalanced ones and provides an additional performance measure that reflects the balance within the output mix of a DMU.

4.2.2 Two-step procedure

The procedure to evaluate the relative balance is comprised of two main steps: determination of the acceptable output mix region and calculation of each DMU's balance score. These steps are explained below for the case in which the crucial performance criteria are represented by the outputs. Consequently, the DMUs' inputs are not taken into consideration.

4.2.2.1 Step 1: Determination of the acceptable output mix region

The first step classifies DMUs as (relatively) balanced or (relatively) unbalanced. For this purpose, the coordinates of an average DMU $AM = (\bar{y}_1,\ldots,\bar{y}_n)$ are determined using the arithmetic mean of $\bar{y}_i$ the n outputs of all given DMUs (y_i^ρ = value of output i for DMU ρ, $i = 1,\ldots,n$; $\rho = 1,\ldots,\pi$). AM possesses a balanced mix in all appointed criteria with regard to the given set of DMUs. As a measure of dispersion for determining the extreme points that will serve as a basis for specifying a multicone, the average absolute deviation from the mean is chosen

$$\bar{d}_i = \frac{\sum_{\rho=1}^{\pi} \left| y_i^\rho - \bar{y}_i \right|}{\pi} \tag{4.1}$$

which is calculated for each of the n outputs, providing an upper and a lower boundary to the maximal admissible variation in each output of AM.

On this basis, intervals $[\bar{y}_i - \bar{d}_i; \bar{y}_i + \bar{d}_i]$ are calculated. The straight lines from the origin through the 2^n combinations $(\bar{y}_1 \pm \bar{d}_1; \bar{y}_2 \pm \bar{d}_2; \ldots; \bar{y}_n \pm \bar{d}_n;)$ determine the limits of the multicone $\mathcal{C}$. This multicone describes the acceptable output mix region for the given DMUs and is defined as

$$\mathcal{C}(F) = \{a\vec{x} : a \in \mathbb{R}, a \geq 0; \vec{x} \in F\} \tag{4.2}$$

It is generated by

$$F := \prod_{i=1}^{n} \left[\bar{y}_i - \bar{d}_i; \bar{y}_i + \bar{d}_i \right], \tag{4.3}$$

where

$$\prod_{\rho=1}^{\pi} \left[a^\rho; b^\rho \right] = \left\{ \vec{x} \in \mathbb{R}^\pi : x^\rho \in \left[a^\rho; b^\rho \right] \right\}. \tag{4.4}$$

Figure 4.2 illustrates Step 1 using a numerical example that will be fully described in Subsection 4.2.3. For 36 pharmacy stores, two outputs are considered: the absolute increase of the turnover (TURN) and the number of the purchases of over-the-counter (OTC) products. The shaded region represents the multicone $\mathcal{C}$, i.e., the acceptable output mix region, which envelops the DMUs whose performance criteria, here represented by the outputs, are achieved on a (relative) similar level. Hence, the DMUs outside the multicone – A7, A15, A17, A22, A23, A24, A27, and A28 – do not achieve their performance criteria in an acceptable proportion. However, to use this information for rewarding or penalizing the DMUs is only possible if the extent to which each DMU avoids concentration on only some of the crucial performance criteria is evaluated. This is described in the next step of the procedure.

Fig. 4.2: Construction of the acceptable output mix region $\mathcal{C}$ for 36 DMUs and two outputs

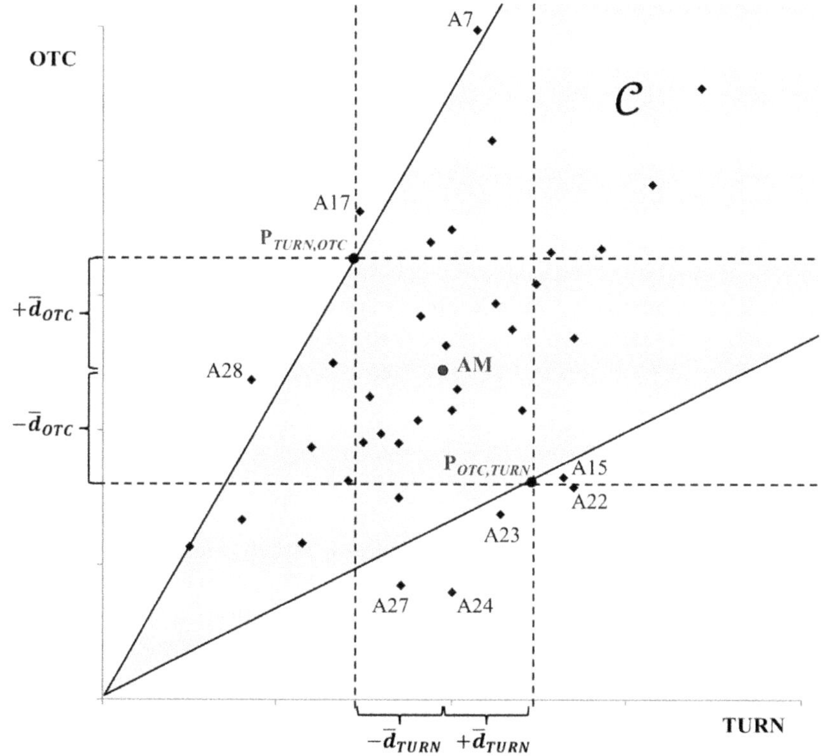

4.2.2.2 Step 2: Calculation of the balance score

In this step, the degree of balance of the DMU is quantified. For all DMUs located inside and on the boundary of the multicone $\mathcal{C}$, the balance score is defined as 100%; in relation to all DMUs under evaluation, these DMUs are characterized by an acceptable mix of the outputs. All other DMUs will be characterized by an individual balance score lower than 100%. The formula $1-\beta^0$ reflects to what extent the extreme pattern of the output values of an unbalanced DMU differs from that of a reference point located on the boundary of the multicone.

Various options are worth considering in determining such a reference point. A first intuitive thought may be to choose a point representing an improvement in as few outputs as possible without worsening the others. However, this allows extreme solutions in the sense that the DMU would have to be able to drastically increment its output production to become balanced. Here, a trade-off between the outputs is considered instead. This trade-off takes into account the practical efforts needed to turn an unbalanced output mix into a balanced one. The focus is on the relative variation w_i^o of each output i required to project an unbalanced DMU onto the boundary of $\mathcal{C}$.

The goal function of the balance model (4.6) determines the values of w_i^o such that the maximum relative variation necessary to obtain a balanced output mix is minimal. In this function, β^0 represents the degree of balance concerning the output mix of DMU^o, with higher scores being more desirable. The ratio

$$\frac{y_t^o\left(1+w_t^o\right)}{y_i^o\left(1+w_i^o\right)} \tag{4.5}$$

represents the slope of the line from the origin through the reference point. A restriction ensures that this ratio is bounded by $up_{i,t}$ and $low_{i,t}$, where $up_{i,t}$ represents the slope of the upper boundary of the intersection $\mathcal{C}_{i,t}$ of the multicone and the $y_i y_t$-plane, and $low_{i,t}$ represents the slope of the lower boundary. Because DMUs with a zero value in at least one of the outputs cannot make a relative improvement, a further restriction replaces all $y_i^o = 0$ with $y_i^o \geq \varepsilon_1$, where ε_1 is a small positive value. Finally, it is not permissible for the output values to be negative, implying that $w_i^o \geq -1+\varepsilon_2$, where ε_2 is a small positive value. The following linear program summarizes the calculation of the balance score β^0 for DMU^o:

$$\begin{aligned}
&\beta^o = 1-\min\left[\max\left(\left|w_i^o\right|,\ldots,\left|w_n^o\right|\right)\right] \\
\text{s.t.}\quad & up_{i,t} \geq \frac{y_t^o\left(1+w_t^o\right)}{y_i^o\left(1+w_i^o\right)} \geq low_{i,t} \qquad i,t=1,\ldots,n \wedge (i<t) \\
& y_i^o \geq \varepsilon_1 \geq 0 \qquad i=1,\ldots,n \\
& w_i^o \geq -1+\varepsilon_2 \qquad i=1,\ldots,n
\end{aligned} \tag{4.6}$$

The proposed model is units-invariant. This is proved by fixing an output i and rescaling y_i^o by a scalar $q>0$. The corresponding second constraint of model (4.6) becomes:

$$up_{i,t} = \frac{\left(\bar{y}_t + \bar{d}_t\right)}{q\left(\bar{y}_i - \bar{d}_i\right)} \geq \frac{y_t\left(1+w_t^o\right)}{qy_i\left(1+w_i^o\right)} \geq \frac{\left(\bar{y}_t - \bar{d}_t\right)}{q\left(\bar{y}_i + \bar{d}_i\right)} = low_{i,t} \quad i,t = 1,\ldots,n \wedge (i<t) \qquad (4.7)$$

This modification does not affect the determination of the corresponding values of w_t^o and w_i^o, and thus generates an optimal solution equivalent to that determined by the original linear program (4.6).

Fig. 4.3: Projection of the unbalanced DMUs onto $\mathcal{C}$

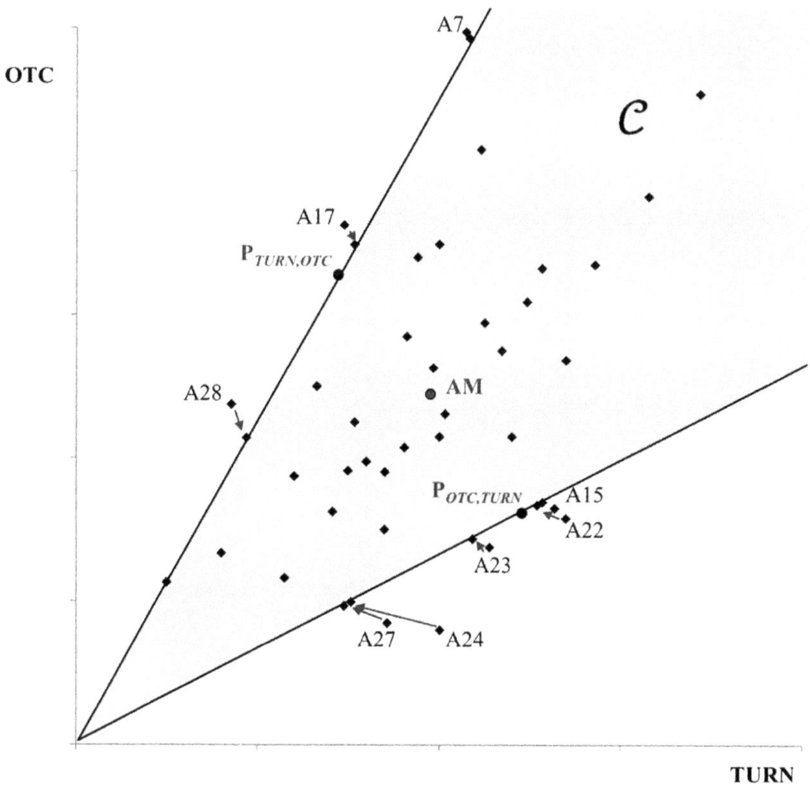

Figure 4.3 illustrates Step 2, showing how the pharmacy stores A7, A15, A17, A22, A23, A24, A27, and A28 are projected onto the surface of the multicone $\mathcal{C}$. A deeper discussion based on an extended numerical example is presented in the next subsection.

Tab. 4.1: Extended data set of the 36 pharmacy stores

DMU	Inputs		Outputs		
	WH	SQM	TURN	OTC	RES
A1	12,005	107	9,938	154,500	20,000
A2	12,120	110	11,439	167,450	18,490
A3	14,357	112	13,726	227,050	16,930
A4	13,895	130	8,917	207,650	17,130
A5	11,408	136	12,604	191,250	16,140
A6	12,847	134	10,802	134,250	18,000
A7	15,342	105	8,576	248,750	15,650
A8	9,743	87	9,006	147,200	15,500
A9	11,986	90	10,273	166,250	15,160
A10	11,387	125	7,511	170,000	14,180
A11	10,563	78	9,381	137,500	12,550
A12	11,951	100	9,612	107,500	13,860
A13	7,995	118	7,990	174,550	10,920
A14	10,468	115	7,862	131,450	10,380
A15	9,024	70	10,561	82,500	14,840
A16	12,648	88	7,281	142,500	10,110
A17	14,763	122	5,899	181,250	12,230
A18	12,154	75	5,981	95,600	17,860
A19	8,675	76	8,000	107,500	9,460
A20	13,040	98	5,295	125,000	12,390
A21	14,057	114	8,120	115,400	6,940
A22	9,687	84	10,799	78,900	9,870
A23	7,572	90	9,113	68,850	5,800
A24	8,912	82	8,014	40,000	7,830
A25	12,046	135	6,388	98,750	14,670
A26	9,080	86	6,791	75,000	14,510
A27	10,099	66	6,849	42,500	14,350
A28	13,003	85	3,424	118,750	11,580
A29	7,541	68	5,640	81,250	14,020
A30	10,759	92	3,211	66,750	13,730
A31	11,703	128	7,223	103,750	10,920
A32	13,854	122	6,130	112,500	8,480
A33	9,670	80	2,010	56,700	12,840
A34	6,241	65	6,797	95,250	2,060
A35	7,846	93	4,806	93,750	11,250
A36	8,694	72	4,598	58,100	10,050

4.2.3 Illustrative example

For clarity and consistency, the example of 36 shopping center pharmacies presented in the previous chapter is continued. As Table 4.1 shows, the pharmacies were analyzed based on three outputs as goal indicators measuring the achievement level of customer-oriented goals to be pursued by the stores[261]:

- TURN: absolute increase of the turnover (in Tsd. €);
- OTC: number of the purchases of over-the-counter products;
- RES: number of the responses to an annual competition in each store's catchment area.

Two inputs are taken into account, representing the means required to achieve these goals:

- WH: worked hours;
- SQM: store square meters.

The statistics characterizing the data set are given in Table 4.2. A closer look at these statistics reveals a wide dispersion in the achievement levels of the outputs: the average absolute deviation from the mean is even higher than the minimum values of the outputs. This indicates that there might be some DMUs which performance criteria values are not achieved on a similar level – relative to the respective values of the other DMUs. Therefore, the evaluation of relative efficiency of the pharmacy stores concerning the achievement of customer-oriented goals is supported by a balance analysis to avoid the overestimation of efficiency through neglecting of some crucial performance criteria.

Tab. 4.2: Descriptive statistics

	Inputs		Outputs		
	WH	**SQM**	**TURN**	**OTC**	**RES**
Minimum	6,241.00	65.00	2,010.00	40,000.00	2,060.00
Maximum	15,342.00	136.00	13,726.00	248,750.00	20,000.00
Mean	11,031.50	98.28	7,793.53	122,386.10	12,796.67
Mean absolute deviation	1,942.61	18.92	2,036.44	41,579.01	3,040.19

261 For more information about the scope of the assessment and the respective example, see Subsection 3.3.3.

Tab. 4.3: Results of the balance model (iteration 1)

DMU	Balance score	Relative output variations		
		w^o_{TURN}	w^o_{OTC}	w^o_{RES}
A1	1	0	0	0
A2	1	0	0	0
A3	1	0	0	0
A4	1	0	0	0
A5	1	0	0	0
A6	1	0	0	0
A7	0.99	0.01	–0.01	0
A8	1	0	0	0
A9	1	0	0	0
A10	1	0	0	0
A11	1	0	0	0
A12	1	0	0	0
A13	1	0	0	0
A14	1	0	0	0
A15	0.97	–0.03	0.03	0
A16	1	0	0	0
A17	0.96	0.04	–0.04	0
A18	0.96	0.04	0	–0.04
A19	1	0	0	0
A20	1	0	0	0
A21	0.93	–0.07	0	0.07
A22	0.94	–0.06	0.06	0.02
A23	0.78	–0.22	–0.15	0.22
A24	0.76	–0.24	0.24	0.07
A25	1	0	0	0
A26	1	0	0	0
A27	0.73	–0.08	0.27	–0.27
A28	0.90	0.10	–0.09	–0.10
A29	1	0	0	0
A30	0.78	0.22	0.02	–0.22
A31	1	0	0	0
A32	1	0	0	0
A33	0.60	0.39	0.40	–0.40
A34	0.30	–0.70	–0.38	0.70
A35	1	0	0	0
A36	1	0	0	0

The main numerical results calculated by running the balance model are presented in Table 4.3. The DMUs A7, A15, A17, A18, A21, A22, A23, A24, A27, A28, A30, A33, and A34 are the ones which do not attain all outputs in an acceptable proportion. For example, DMU A23 has a balance score β^{A23} of 0.78 since it requires a minimal output variation of 22% to be projected onto $\mathcal{C}$; an increase of 22% on RES ($w_{RES}^{A23} = 0.22$) is necessary, while a reduction of 22% on TURN ($w_{TURN}^{A23} = -0.22$) and 15% on OTC ($w_{OTC}^{A23} = -0.15$) is allowed. These moderate w_i^0 values indicate that the reference point of A23 is a suitable (balance) benchmark.

In cases, where at least one output is neglected, i.e. produced on a low level or not produced at all, the respective DMU will be penalized by a low balance score – in the extreme case, by a balance score of 0. In this example, A34 represents a DMU whose low level of customer responses (RES) indicates that it seems not to be interested in pursuing the goal of increasing its popularity. The corresponding result of the linear program (4.6) is the balance score β^{A34}=0.30, with $w_{TURN}^{A34} = -0.70$, $w_{OTC}^{A34} = -0.38$, and $w_{RES}^{A34} = 0.70$. These w_i^0 values project A34 onto the tip surface of the multicone, which hardly can be understood as a benchmark. In contrast, it indicates that A34 follows a different business strategy and therefore may not be comparable to the other DMUs.

The results of the efficiency analysis, i.e., the efficiency scores and the matrix of the benchmarks obtained from the output-oriented CCR (CCR-O) model, are given in Table 4.4. Ten DMUs A3, A5, A7, A8, A13, A15, A18, A23, A29, and A34 are characterized as efficient, or more precisely as strong efficient. However, five of these DMUs have an unbalanced output mix, i.e. they possibly benefit from the flexibility of the weights leading to the exclusion of those input and/or output criteria in which the performance of a DMU is weak. Moreover, the matrix of the benchmarks discloses that unbalanced DMUs – highlighted by grey – are frequently used as benchmarks for the inefficient DMUs. These results show that a balance analysis can have various and substantial implications for management. A structured overview of these implications is presented in the next section.

Tab. 4.4: Results of the CCR-O model

DMU	Efficiency score	Benchmarks									
		A3	A5	A7	A8	A13	A15	A18	A23	A29	A34
A1	0.98				1		1	1		1	
A2	0.96		1		1	1	1				
A3	1	1									
A4	0.90			1	1	1					
A5	1		1								
A6	0.85					1	1			1	
A7	1			1							
A8	1				1						
A9	0.99	1		1	1		1				
A10	0.84			1	1	1					
A11	0.96	1		1			1				
A12	0.75	1	1		1		1				
A13	1					1					
A14	0.73	1	1		1	1					
A15	1						1				
A16	0.76	1		1			1				
A17	0.73			1	1	1					
A18	1							1			
A19	0.88	1	1				1				1
A20	0.70			1	1			1			
A21	0.57	1					1				1
A22	0.95						1		1		
A23	1								1		
A24	0.76						1		1		
A25	0.68					1	1			1	
A26	0.88						1			1	
A27	0.96						1	1			
A28	0.73			1				1			
A29	1									1	
A30	0.71							1		1	
A31	0.60				1	1	1			1	
A32	0.49	1		1	1	1					
A33	0.76							1		1	
A34	1										1
A35	0.84					1				1	
A36	0.66				1		1	1		1	
Sum		**9**	**5**	**10**	**13**	**11**	**17**	**8**	**3**	**10**	**3**

4.3 Managerial implications

Complementing a usual DEA efficiency analysis with the proposed balance analysis allows the classification of efficient as well as inefficient DMUs into those which are balanced and those which are not. Table 4.5 shows the resulting extended DMU structure in the numerical example and depicts its benefits from a managerial point of view.

Tab. 4.5: Extended DMU structure and its benefits

DMUs to be classified as	**Pharmacy stores**	**Benefit for performance assessment**	**Benefit for performance control**
Efficient and balanced	A3, A5, A8, A13, A29	Ranking according to balance scores	Reduced set of benchmarks
Efficient but unbalanced	A7, A15, A18, A23, A34		
Inefficient but balanced	A1, A2, A4, A6, A9, A10, A11, A12, A14, A16, A19, A20, A25, A26, A31, A32, A35, A36	Indication of DMUs with a further performance deficit	Indication of priorities concerning measures for efficiency improvement
Inefficient and unbalanced	A17, A21, A22, A24, A27, A28, A30, A33		

The complementary information provided by the balance analysis allows a broader performance assessment to be conducted. Concerning the set of efficient DMUs, it seems appropriate to use their balance score for a (partial) ranking. In the example, A3, A5, A8, A13, and A29 are equal in the first position with a balance score of 1, followed by A7 ($\beta^{A7} = 0.99$), A15 ($\beta^{A15} = 0.97$), A18 ($\beta^{A18} = 0.96$), A23 ($\beta^{A23} = 0.78$), and A34 ($\beta^{A34} = 0.30$). A further analysis of the results, especially in the case of extremely unbalanced DMUs, is essential. For example, it may be found that such a DMU is following a significantly different business strategy and therefore should be excluded from the set of comparable DMUs.

A ranking concept for the set of inefficient DMUs would require a trade-off between the efficiency scores and the balance scores. Here, it is assumed that this does not correspond to the relationship between these measures. Regarding the pharmacy stores, it is considered to be negligible that a store is balanced if it has a low efficiency score (e.g., A21). With an increasing efficiency, however, the balance of the respective stores (e.g., A22) starts gaining importance. This characterizes the balance score as a kind of secondary measure. On one hand, efficiency deficits cannot be compensated by a high balance score. On the other hand, balance deficits let the performance expressed by a high efficiency score appear in another light. Again, the question arises of whether a different business strategy is being pursued.

The results of the balance assessment should also be studied in regard to their impact on performance control. Measures intended to increase the performance of inefficient DMUs should be reconsidered. If they are balanced, measures promising to keep this property should be chosen; if they are unbalanced, measures attempting to raise the achievement of outputs neglected so far should be prioritized. The w_i^0 values indicate the extent of the necessary output increases and allowed output decreases, if inevitable.

From the perspective of the efficient DMUs, management should be cautious about awarding (also) those which are unbalanced. By concentrating efforts on only a few performance criteria, this would create the usually undesired incentive to neglect the unit balance in favor of raising the unit efficiency. In this context, analysts should especially reconsider using efficient but unbalanced DMUs as benchmarks. An inefficient DMU with such a benchmark would be urged to pursue a questionable performance strategy. For example, the CCR-O model designates A34 as a benchmark for A19 and A21 (see Table 4.4), although Table 4.1 shows that A34 is more concentrated on the achievement of two outputs (TURN and OTC) neglecting the low level of the third output (RES). The pharmacy stores A19 and A21, on the contrary, are focusing on all three outputs.

Against this background, it is suggested to exclude efficient but unbalanced DMUs from the set of possible benchmarks. This can be achieved by adding the last constraint to the envelopment form of the CCR-O model:

$$\begin{aligned}
&\max\ \varphi_*^o \\
\text{s.t.}\quad & x_j^o \geq \sum_{\rho=1}^{\pi} \lambda^\rho x_j^\rho \qquad j = 1,\ldots m \\
& \varphi_*^o y_i^o \leq \sum_{\rho=1}^{\pi} \lambda^\rho y_i^\rho \qquad i = 1,\ldots,n \\
& \lambda^\rho \geq 0 \qquad \rho = 1,\ldots,\pi \\
& \lambda^\rho = 0 \qquad \rho \neq o,\ \beta^\rho \neq 1
\end{aligned} \tag{4.8}$$

This model still allows all efficient DMUs to serve as their own benchmarks, while eliminating unbalanced DMUs from the efficiency frontier relevant for the inefficient DMUs. Consequently, the reference sets of the inefficient DMUs only contain balanced DMUs, and the modified efficiency frontier leads to improvements of the efficiency scores of inefficient DMUs. These improvements result from the fact that extreme targets set by the unbalanced benchmarks are replaced with more moderate targets.

In special cases, it may even happen that an inefficient DMU turns into an efficient one because it becomes part of the new frontier. Table 4.6 presents the results of the modified CCR-O model for the numerical example introduced in Subsection 4.2.3.

Tab. 4.6: Results of the modified CCR-O model (iteration 1)

DMU	Efficiency score	Benchmarks													
		A3	A5	A7	A8	A9	A11	A13	A15	A18	A22	A23	A27	A29	A34
A1	0.98				1	1	1							1	
A2	1	1	1		1										
A3	1	1													
A4	0.90	1			1			1							
A5	1		1												
A6	0.88		1		1									1	
A7	1			1											
A8	1				1										
A9	1					1									
A10	0.85	1			1			1							
A11	1						1								
A12	0.85	1	1		1										
A13	1							1							
A14	0.73	1	1		1			1							
A15	1								1						
A16	0.80	1													
A17	0.76	1						1							
A18	1									1					
A19	0.93	1	1												
A20	0.73				1	1									
A21	0.60	1	1												
A22	1										1				
A23	1											1			
A24	0.89	1	1												
A25	0.68		1					1						1	
A26	0.91		1		1									1	
A27	1												1		
A28	0.79				1	1									
A29	1													1	
A30	0.72													1	
A31	0.61		1		1									1	
A32	0.49	1			1			1							
A33	0.78													1	
A34	1														1
A35	0.84							1						1	
A36	0.71					1								1	
Sum		12	11	1	13	5	2	8	1	1	1	1	1	10	1

The ten pharmacy stores originally identified as efficient keep this property. While the balanced DMUs A3, A5, A8, A13, and A29 also continue to serve as benchmarks for other stores, the unbalanced DMUs A7, A15, A18, A23, and A34 are eliminated from the reference set of other stores. This increases the efficiency score of those inefficient stores which formerly had unbalanced DMUs in their reference sets. For example, the new benchmarks of A2 are A3, A5, and A8, raising its efficiency score from 0.96 to 1. As special cases, A9, A11, A22, and A27 are now characterized as efficient. However, only A9 and A11 become appropriate benchmarks for other DMUs, because they are balanced.

On the basis of the data analysis in the initial evaluation, designated as first iteration, the central management of the pharmacy stores should take decisions according to the respective results. If a store is unbalanced, a change of its output mix in reference to the determined reference point on the surface of the multicone is required. Hence, the DMUs A7, A15, A17, A18, A21, A22, A23, A24, A27, A28, A30, A33, and A34 should adjust their output values according to the proposed w_i^0 values in Table 4.3. Such changes in the output mix may induce a negative change in the efficiency scores in the subsequent evaluation, indicating the overestimation of the actual performance of the DMUs.

Assuming that the balanced DMUs keep their balanced output mix, the efficiency and balance of pharmacy stores can be evaluated a second time. The DMUs A15 ($\beta^{A15} = 0.99$), A22 ($\beta^{A22} = 0.99$), A23 ($\beta^{A23} = 0.99$), A24 ($\beta^{A24} = 0.99$), and A27 ($\beta^{A27} = 0.98$) still remain unbalanced, while A7, A17, A18, A21, A28, A30, A33, and A34 become balanced. The results of the second iteration of the modified CCR-O model are given in Table 4.7.

The results in Table 4.7 show a decrease of the efficiency scores of unbalanced DMUs due to the changes in the output mix. Therefore, the number of efficient DMUs is reduced to 12 in comparison to 15 DMUs in the first iteration (see Table 4.6). As a consequence, the number of benchmarks reduces as well, affecting the reference set of inefficient DMUs. The efficiency scores of inefficient but balanced DMUs with such a change in the reference set decrease, but not below the efficiency level calculated by CCR-O model without the modification (see Table 4.4). For example, the efficiency score of DMU A36 decreases to 0.70, which is lower than the efficiency score of A36 in iteration 1 (see Table 4.6), but higher than the efficiency score provided in Table 4.4. The rest of the balanced DMUs keep their efficiency scores from the previous period.

Tab. 4.7: Results of the modified CCR-O model (iteration 2)

DMU	Efficiency score	Benchmarks										
		A3	A5	A7	A8	A9	A11	A13	A15	A18	A22	A29
A1	0.98				1					1		1
A2	1	1	1		1							
A3	1	1										
A4	0.90			1	1			1				
A5	1		1									
A6	0.88		1		1							1
A7	1			1								
A8	1				1							
A9	1					1						
A10	0.85			1	1			1				
A11	1						1					
A12	0.85	1	1		1							
A13	1							1				
A14	0.73	1	1		1			1				
A15	1								1			
A16	0.76	1		1						1		
A17	0.71			1	1			1				
A18	1									1		
A19	0.93	1	1									
A20	0.70			1	1					1		
A21	0.55	1	1									
A22	1										1	
A23	0.85	1	1									
A24	0.69	1	1		1							
A25	0.68		1					1				1
A26	0.91		1		1							1
A27	0.73					1				1		
A28	0.67			1						1		
A29	1											1
A30	0.57				1					1		1
A31	0.61		1		1							1
A32	0.49	1		1	1			1				
A33	0.55			1	1					1		
A34	0.50			1				1				
A35	0.84							1				1
A36	0.70				1					1		1
Sum		**10**	**12**	**10**	**17**	**1**	**2**	**9**	**1**	**9**	**1**	**9**

This iterative process of balance and efficiency analysis could be continued. Assuming that the output mix of balanced DMUs remains constant, the unbalanced DMUs will turn into balanced ones after further iterations will be conducted.

4.4 Brief review of recent model developments

A series of alternative models – each with its own characteristics – can be developed by modifying the procedure to calculate the balance score. The first alternative model has been introduced by Dyckhoff et al. (2013). Their balance model differs from that presented in Section 4.2 in two main aspects:[262]

- the integration of the evaluation of balance into the methodology of DEA and
- the alternative determination of the balance score.

The postulate that the evaluation of balance should be part of the usual DEA methodology affects the determination of the projections of unbalanced points onto the set of balanced DMUs. Dyckhoff et al. (2013) proposed the following assumptions: (i) the reference point of an unbalanced DMU should be weakly dominated concerning its outputs by the corresponding unbalanced DMU; (ii) this reference point should reach the highest possible efficiency score. The degree of balance is then calculated as the ratio of two efficiency scores – the one of the balanced reference point, called balanced efficiency, and the one of the initial (unbalanced) DMU.[263]

The main emphasis of Dyckhoff et al.'s (2013) publication was placed on an elaborate illustration of the new approach to the data on the research performance of German business schools taken from the ranking of the Centre of Higher Education.[264] In contrast, the recent paper of Dyckhoff et al. (2015) focused on a deeper mathematical explanation and a general formulation of the model for evaluating balance and the corresponding (opposite) specialization degree of DMUs. Furthermore, in the publication from 2015, the authors reviewed and compared in detail their approach to the one presented in Section 4.2 and published as Ahn et al. (2012a), using the same numerical example. The following aspects have been mentioned by Dyckhoff et al. (2015) as the main different characteristics of their approach:[265]

---

262 Cf. Dyckhoff et al. (2013), p. 42.
263 Cf. Dyckhoff et al. (2013), p. 44.
264 Cf. Dyckhoff et al. (2015), p. 198.
265 Cf. Dyckhoff et al. (2015), p. 210.

- *DEA-integrated evaluation of balance and specialization*
  The assumption of a trade-off between efficiency (or effectiveness) and balance (or specialization, respectively) leads to a systematic integration of the evaluation of balance as well as specialization into the methodology of DEA. Firstly, the balance score β is defined as the quotient of the associated efficiency scores as mentioned by Dyckhoff et al. (2013) – the corresponding (opposite) specialization score is calculated as $1 - \beta$. Secondly, the determination of the balanced reference point is included in the DEA methodology through the modification of basic DEA models.
- *No compensatory balance and specialization measure*
  The projection of unbalanced (specialized) DMUs onto balanced reference points does not allow for any kind of substitution between the outputs as crucial performance criteria.
- *Optional convex polyhedral balance cones*
  The balance output mix region in the output space is defined as the intersection of the (chosen) data envelopment, i.e. the predefined set of balanced output mixes, with any convex polyhedral cone. Therefore, various balance output mix regions can be easily determined according to the alteration of the (chosen) data envelopment.
- *Guarantee of feasible solutions*
  Due to the "free disposal"[266] assumption of basic DEA models, the non-compensative character of the projection of unbalanced (specialized) DMUs onto the balanced output mix region guarantees the existence of a balanced reference point with the corresponding data envelopment.

The brief review reveals that the idea of evaluating the relative balance of DMUs as an additional performance indicator is welcomed by other authors as well. However, due to certain variations in the assumptions or characteristics of the approaches, the calculation of the balance score substantially differs between them. As a consequence, the respective results, i.e. balance scores, are not directly comparable across these approaches. The decision concerning the selection of the appropriate approach for the specific application is therefore up to the decision makers.

266 According to Cooper et al. (2007b), p. 73, „free disposal" means that disposing of excess slacks in inputs and/or outputs is not associated with any cost. Therefore, the non-zero slacks that may occur in inputs and/or outputs are not captured by the measure, i.e. they are ignored.

4.5 Conclusion and discussion

This chapter has addressed the managerial problem of a possible neglect of performance criteria within the performance evaluation. In DEA, this problem occurs due to the flexibility in weight determination, which permits zero-value weights, excluding the respective performance criteria from the evaluation. To address the problem of zero-value weights, a research area called value judgment has originated in DEA literature. A brief review of the value judgment literature has not revealed an overall solution to the limitations stemming from flexible weights. Besides the mathematical limitations of the approaches incorporating value judgments in DEA, the requirement of additional subjective information concerning the respective evaluation case – leading thereby to an increase of managers' workload – has been detected as a considerable disadvantage. To avoid the problems resulting from flexible weights without the need for further value judgments, an innovative approach that complements the two common aspects of performance – efficiency and effectiveness – with a measure of balance has been developed. The balance score evaluates the extent to which the crucial performance criteria, here represented by the outputs, are achieved in an acceptable proportion. Based on a typical data set for a DEA application, it allows for a higher discrimination within both efficient and inefficient DMUs by quantifying their relative balance.

A thorough evaluation of the achieved balance scores provides insights that substantially influence the DEA efficiency analysis. Concerning performance assessment, the explanation for a DMU's low balance score could be that the DMU pursues a different business strategy and therefore is not comparable to the other DMUs. Concerning performance control, the management may agree that efficient but unbalanced DMUs are not adequate benchmarks for inefficient DMUs. In this case, a modified CCR-O model has been proposed, which endogenously calculates the balance score and eliminates the undesirable benchmarks. Furthermore, an iterative process of balance and efficiency analysis has been illustrated by continuing the example of shopping center pharmacies presented in Chapter 3. This iterative process determines adequate efficiency scores and supports the unbalanced DMUs on their way to get balanced.

The presented approach for evaluating the relative balance of DMUs has already been published as Ahn et al. (2012a) inducing thereby the development of alternative models to evaluate the balance of DMUs. A brief review of the recent model developments has revealed the acceptance of relative balance as a complementary performance measure. However, it has also detected substantial differences in the assumptions which underlie the approaches, e.g. concerning the incorporation of additional information, compensatory vs. non-compensatory property of balance score, as well as the incorporation vs. non-integration of balance into the methodology of DEA. Although, these approaches can be subsumed under the family of models for evaluating the relative balance, their results, i.e. the respective balance scores, are not comparable. This implies that the suitability of an approach depends on the match of the characteristics between the approach and the particular evaluation case.

From a managerial point of view, the approach for evaluating the relative balance of DMUs presented within this chapter is aimed to meet the requirements on model design provided by Little (2004). Its strengths and limitations referring to these requirements are presented in Table 4.8.

Tab. 4.8: Strengths and limitations of the presented approach for evaluating the relative balance of DMUs

Managerial requirements on model design[267]	Strengths	Limitations
Simple	Comprehensible quantification of the balance scores No need for incorporation of additional information (e.g. weights)	Understanding of mathematical background required
Robust	Desirable sensitivity to the distribution of the DMUs Replicable measure of relative balance based on the comparison with a balanced output mix	Not suitable in case of outputs with multimodal distributions
Easy to control	Based on a typical data set for a DEA application	
Adaptive	Units-invariant approach	Potential danger of mixing short with long term measures as well as indices with volume measures
Complete on important issues	Simultaneous consideration of crucial performance criteria, here outputs	Focus only on the outputs, neglect of the inputs
Easy to communicate with	Possibility of iterative evaluation of balance and efficiency Higher discrimination within efficient and inefficient DMUs	Iterative evaluation implies the output mix of balanced DMUs to remain constant

267 Cf. Little (2004), pp. 1843–1844.

Since the emphasis through the development of the approach consists particularly in keeping the managers' workload to a minimum, the approach ensures a comprehensible quantification of the balance scores without the need of any additional information. Therefore, it can be characterized as simple. However, there is still the need for decision makers to understand the mathematical background of the proposed approach and of DEA methodology in general.

A further desirable feature of the approach is the generation of a multicone that is sensitive to the distribution of the DMUs, changing its position and amplitude according to the output values of the DMUs. In this way, the approach is able to depict an appropriate basis for the creation of a balanced output mix, resulting in a robust calculation of the balances scores. However, special cases of outputs with multimodal distributions require an extension of the approach, because a single multicone is not able to reflect the data structure of the corresponding DMUs. A solution could be to classify the DMUs into different clusters and to determine a multicone for each of these clusters.

As already mentioned, one of the most important strengths of the approach is its use of already prepared or selected information, e.g. the data set needed for a usual DEA application. This feature enables the approach to be easy to control. Furthermore, the characteristic of being units-invariant makes the approach adaptive to new information. On the other side, the well-known problem of mixing short-term with long-term measures as well as indices with volume measures mentioned by Dyson et al. (2001) remains unsolved. This concerns especially the assumption that a DMU is generally able to substitute its outputs, one for the other in line with its capacity. The proposed relative variations w_i^0 required to turn an unbalanced DMU to a balance one have to be applied with cautiousness due to the possibility that some of them may not always be fully attainable.

According to the underlying interpretation of outputs as indicators of fundamental goals, the attention within the proposed approach is focused only on the balance of the output mix – as crucial performance criteria. From such a point of view, the approach is complete on important issues. However, cases are imaginable, which do not correspond to the assumption that the main goals of the DMUs are reflected only by the outputs. For other input/output subsets, it is appropriate to derive specific balance scores, e.g. by a modification of the calculation procedure.

Finally, the main benefit of the presented approach is its complementary provided information that substantially influences the DEA efficiency analysis. One the one hand, it allows a higher discrimination within efficient and inefficient DMUs resulting in consequences concerning not only the performance assessment, e.g. the comparability of the DMUs in the sample, but also the performance control, e.g. questionable use of efficient but unbalanced DMUs as benchmarks. On the other hand, its easily explainable results suggest an iterative evaluation of balance and efficiency of the DMUs in order to support the unbalanced DMUs with improvement advices to turn them into balanced ones. However, the proposed iterative evaluation process is only valid as long as the output mix of balanced DMUs remains constant. In practice, this is solely possible over a short period of time. The more time passes, the more likely it is that the output mix of the DMUs – independent of balanced or unbalanced – may change, meaning that the current iterative evaluation process should no longer be continued. Instead, a new evaluation process of balance for the whole sample is necessary to get adequate results. Thus, the evaluation of relative balance – like efficiency and effectiveness – is an ongoing process.

5 Evaluation of relative efficiency[268,269]

Performance evaluation focuses on the determination whether an organization is successful in aligning its performance in accordance with its goals. Therefore, it requires the explicit consideration of organizational goals within the evaluation process. The traditional DEA methodology implies that performance criteria used for the *efficiency* evaluation are related to the goals of an organization, but does not specify how these goals are taken into account. Furthermore, the determination of performance criteria as well as their characterization as inputs or outputs is also not outlined in detail. In light of the fact that DEA results rely heavily on the selected set of inputs and outputs as performance criteria,[270] this lack of transparency raises doubts as to the plausibility of the results, which in turn negatively affect the acceptance and use of DEA in practice.

As a consequence, DEA has to be seen in a broader context, supporting the initial proposal of Dyckhoff & Allen (2001), explicated by Dyckhoff & Ahn (2010), for a generalized DEA (GDEA). Their approach ensures that organizational goals are included in the efficiency evaluation by adjusting the traditional DEA models to the evaluation of goals instead of inputs and outputs. Inputs and outputs, as performance criteria, are thereby used to operationalize the goals. Despite this potential of GDEA to systematically link goals and corresponding performance criteria, the specification of a valid GDEA model for the respective application case can be challenging, requiring methodical support for the users, such as decision makers. Therefore, the aim of this chapter is to introduce a stepwise approach that assists the decision makers in developing an appropriate GDEA model based on the determination of suitable and measurable performance criteria from goal-oriented perspective. In doing so, the basic concepts of multi-criteria decision

268 The content described in this chapter has been partially published by Afsharian, M., Ahn, H., & Neumann, L. (2016): Generalized DEA: an approach for supporting input/output factor determination in DEA, in: Benchmarking: An International Journal 23(7), pp. 1892–1909. The content contains some minor changes in comparison with the published version.

269 Some parts of the content described in this chapter has been published by Ahn, H., & Neumann, L. (2014): Measuring the efficiency of pharmacy stores with respect to customer retention, in D.V. Vasilenko, G.L. Bagiev, W. Fritz, O.U. Yuldasheva, & J.N. Solovjova (Eds.): Marketing Competencies and Business Performance: Proceedings of the International Research Symposium (pp. 46–51). May 2014, St. Petersburg, Russia.

270 Wagner & Shimshak (2007), pp. 57–58.

making (MCDM)[271] for determining, structuring, and operationalizing objectives are utilized in the GDEA context.

The chapter is organized as follows. In Section 5.1, a rethinking of the DEA efficiency evaluation is proposed. By describing the problems in the performance criteria determination, which result from the restricted focus of DEA efficiency evaluation on the transformation process of inputs into outputs, the need to expand the DEA methodology by GDEA is revealed. A brief review of GDEA emphasizes, however, the importance of a methodical support for the decision makers that facilitates the implementation of GDEA. Therefore, Section 5.2 introduces a goal-oriented approach for determining performance criteria required for GDEA efficiency evaluation, referring to the example of shopping center pharmacies. The insights gained within this stepwise approach confirm that GDEA can cope with outlined problems in the performance criteria determination, addressing their cause. Corresponding managerial implications are explained in Section 5.3. Finally, Section 5.4 concludes the chapter and discusses the characteristics of the introduced approach with regard to the fulfillment of requirements on model design from the managerial point of view.

5.1 Rethinking DEA efficiency evaluation

5.1.1 Problems in determining performance criteria

Despite the fact that performance criteria provide an essential basis for the performance evaluation, DEA applications usually do not discuss the determination of suitable performance criteria in detail. Moreover, it is not uncommon that the determination is not outlined at all. This is an important pitfall, which may impede the reflection on the appropriateness of DEA results significantly, especially in light of the fact that the following three kinds of problems may occur in any DEA application:

- *Problem of performance criteria selection*: What are the appropriate performance criteria?
  Most of the companies gather information concerning the wide range of their activities. Such information serves as a basis for the initial list of performance criteria potentially relevant for a DEA application. As any criteria which may affect the performance of DMUs under investigation can be taken into con-

271 According to Majumder (2015), p. 35, MCDM deals with "structuring and solving decision and planning problems involving multiple criteria".

sideration, this may result in a considerably large list of criteria.[272] However, the DEA methodology itself is not able to clarify which of these criteria are suitable to reflect the overall performance of the DMUs or a certain aspect of their performance. Thus, the decision makers have to select appropriate criteria, depending on how performance is specified.

- *Problem of dual-role performance criteria*: How to deal with performance criteria which are not clearly inputs or outputs?
  Traditional DEA models presuppose that it is possible to clearly specify which of the performance criteria selected for the evaluation constitute inputs and which outputs.[273] In some situations, however, performance criteria can simultaneously be characterized as both inputs and outputs, e.g. in case of multi-stage production processes.[274] In services process, it is even more difficult to treat the respective services as they may neither be described as inputs nor as outputs. Regarding the example of pharmacy stores and one of their fundamental goals *strengthening of customer retention*, e.g., health counselling services provided by the employees to the customers embodies input and output aspects in uno actu. To deal with such ambiguous criteria, the decision makers have to clearly define the purpose of the evaluation in order to be able to classify them as inputs or outputs.
- *Problem of anti-isotonic or rather undesirable performance criteria*: How to deal with inputs to be maximized and outputs to be minimized?
  DEA methodology is based on the assumption that the performance criteria are isotonic, implying that inputs have to be minimized and outputs to be maximized.[275] However, in some cases, there are anti-isotonic performance criteria that have to be taken into consideration as well. For instance, with regard to the example of pharmacy stores and the evaluation of their efficiency concerning *the strengthening of customer retention*, the telephone hotline hours for health counselling services can be regarded as an input to be maximized. Furthermore, the number of subsequent deliveries of drugs, which were not available in the store when the customers wanted to purchase them, can be regarded as an output to be minimized. Such anti-isotonic performance criteria do not fit into the traditional DEA models and are, therefore, difficult to handle. The DEA literature provides simple heuristics to cope with this pitfall, but the

---

272 Cf. Wagner & Shimshak (2007), p. 58.
273 Cf. Cook & Zhu (2007), p. 692.
274 Cf. Cook et al. (2014), p. 2.
275 Cf. Dyson et al. (2001), p. 250; Cook et al. (2014), p. 2.

selection among these heuristics can have a strong effect on the DEA results for the DMUs under investigation.

A systematic linkage between organizational goals and suitable performance criteria offers one possibility to cope with the outlined problems. For this purpose, Dyckhoff & Allen (2001) and Dyckhoff & Ahn (2010) proposed an extension of the theoretical background of DEA methodology which is briefly presented in the next subsection.

5.1.2 Broadening the methodology of DEA by generalized DEA (GDEA)

The traditional DEA methodology refers to the logic of the classical production theory. DEA uses not only one of the central elements of production theory – the Pareto optimality – for the evaluation of efficiency, but focuses also on inputs and outputs as decisive performance criteria. Additional similarities consist in the assumptions that performance criteria required for the evaluation are usually given, can be clearly specified as inputs or outputs, and are isotonic, meaning that inputs have to be minimized and outputs to be maximized.[276] However, in the context of DEA, these assumptions cause the problems mentioned in the previous subsection and restrict the application potential of DEA to the input-output transformation process without adequately taking organizational goals into account.[277]

Therefore, it seems reasonable to extend the theoretical background of the DEA by the logic of the decision-oriented production theory of Dyckhoff (1992, 2006). This theory incorporates a goal-oriented perspective into the determination of suitable performance criteria, i.e. these criteria – whether inputs or outputs – are selected based upon their ability to reflect the organizational goals. For the first time, Dyckhoff & Allen (2001) applied this theory in the context of DEA and encouraged a rethinking of the respective efficiency evaluation design. They proposed to switch the focus of efficiency evaluation from the solely transformation of inputs into outputs to the accomplishment of respective organizational goals. In doing so, Dyckhoff & Allen (2001) initiated the development of a generalized DEA (GDEA) – extensively explicated by Dyckhoff & Ahn (2010) – which evaluates the efficiency of an organization with regard to its goals to be minimized versus goals to be maximized. Thus, GDEA adjusts the traditional DEA models to the evaluation of goals instead of inputs and outputs. Inputs and outputs, as performance criteria, are only relevant

276 Cf. Dyckhoff & Ahn (2010), pp. 1252–1253.
277 Cf. Ahn & Le (2014), p. 32.

to the extent that they serve for quantifying these goals by means of non-negative value functions.[278]

The explicit reference of GDEA to organizational goals as decisive performance objectives provides an enhanced framework that actively supports managerial decision making in monitoring the performance of an organization in relation to its goals. Furthermore, it establishes a systematic linkage between organizational goals and corresponding performance criteria through the determination of a non-negative value function for each goal. Thereby, it is not important whether the performance criteria chosen to operationalize a particular goal represent inputs or outputs of the underlying production process. This implies that both inputs and outputs can be simultaneously used to adequately reflect the respective organizational goal. In this way, GDEA sheds another light on the problems concerning the determination of suitable performance criteria. To get a better insight into the GDEA methodology and to find out how this approach can contribute to coping with the outlined problems in the performance criteria determination, the next subsection presents a short introduction to GDEA.

5.1.3 Introduction of GDEA

GDEA originates from the intention of Dyckhoff & Allen (2001) to evaluate ecological efficiency based on a more comprehensive preference structure. For this purpose, they suggested

> [...] a generalization of basic DEA models [...] by incorporating a multi-dimensional value function ***f***,[279]

whereby $\xi(\zeta = 1,\ldots,\xi)$ quotes the number of performance objectives considered as relevant for the evaluation. Each performance objective is described as a function of inputs x_j $(j = 1,\ldots,m)$ and outputs y_i $(i = 1,\ldots,n)$ as follows:[280]

$$\boldsymbol{f}: \quad \mathbb{R}^{m+n} \rightarrow \mathbb{R}^{\xi}$$

$$\boldsymbol{f}\begin{pmatrix} \boldsymbol{x} \\ \boldsymbol{y} \end{pmatrix} = \begin{pmatrix} f_1\begin{pmatrix} \boldsymbol{x} \\ \boldsymbol{y} \end{pmatrix} \\ \vdots \\ f_{\xi}\begin{pmatrix} \boldsymbol{x} \\ \boldsymbol{y} \end{pmatrix} \end{pmatrix} \quad (\xi \in \mathrm{N}) \tag{5.1}$$

278 Cf. Ahn & Dyckhoff (2011), p. 3.
279 Dyckhoff & Allen (2001), p. 312.
280 Cf. Dyckhoff & Allen (2001), p. 320.

Dyckhoff & Ahn (2010) explicated that this approach is not only applicable for ecological efficiency, but for any form of functional or rather partial efficiency.

In analogy to traditional DEA, GDEA requires the division of a set of performance objectives considered as relevant for the evaluation into s objectives to be minimized ($g = 1,\ldots,s$ with achieved values k_g) and r objectives to be maximized ($h = 1,\ldots,r$ with achieved values l_h), subsequently called cost and benefit objectives, respectively. The values k_g and l_h of the $s+r = \xi$ performance objectives can be measured according to (5.1) by non-negative value functions $k_g = f_g(\boldsymbol{x}, \boldsymbol{y})$ for the cost objectives and non-negative value functions $l_h = f_{s+h}(\boldsymbol{x}, \boldsymbol{y})$ for the benefit objectives. The multidimensional value function $\boldsymbol{f}$ defined in (5.1) emerges when putting these functions together.[281]

Of particular importance is the case that $\boldsymbol{f}$ as well as the objective functions are linear. In this case, the value k of a cost objective g is equal to the weighted sum of all undesirable (i.e. to be minimized) inputs and outputs τ contributing to g with input/output multipliers $c_{g,\tau}$ as weights. Analogously, the value l of a benefit objective h is equal to the weighted sum of all desirable (i.e. to be maximized) inputs and outputs τ contributing to h with weights in the form of input/output multipliers $e_{h,\tau}$:[282]

$$k_g(\boldsymbol{x}, \boldsymbol{y}) = \sum_{\tau=1}^{m} c_{g,\tau} x_\tau + \sum_{\tau=1}^{n} c_{g,m+\tau} y_\tau$$

and (5.2)

$$l_h(\boldsymbol{x}, \boldsymbol{y}) = \sum_{\tau=1}^{m} e_{h,\tau} x_\tau + \sum_{\tau=1}^{n} e_{h,m+\tau} y_\tau$$

The functional relationships presented in (5.2) provide the basis for the definition of GDEA models. Correspondingly, the goal-oriented efficiency of a DMU is calculated as the ratio of benefit objectives weighted by respective multipliers μ_h to cost objectives weighted by multipliers v_g as follows:[283]

$$\Theta := \frac{L}{K} = \frac{\sum_{h=1}^{r} \mu_h l_h}{\sum_{g=1}^{s} v_g k_g} \tag{5.3}$$

---

281 Cf. Ahn & Dyckhoff (2011), p. 4; Dyckhoff & Ahn (2010), pp. 1259–1260.
282 Cf. Ahn & Dyckhoff (2011), p. 4; Dyckhoff & Ahn (2010), p. 1264.
283 Cf. Dyckhoff & Ahn (2010), p. 1261.

In accordance with the procedure for the development of traditional DEA models, the Charnes-Cooper-Transformation can be applied to (5.3) in order to derive GDEA models. Dyckhoff & Ahn (2010) demonstrated this for the case that the improvement of the cost objectives is emphasized, under the assumption of CRS. By analogy with the CCR-I model of traditional DEA, the resulting model can be named as GCCR-MIN model, where (G) stands for generalized and (MIN) for focusing on the objectives to be minimized.

Since the respective GCCR-MIN as well as GCCR-MAX models are already extensively described and exemplified by Dyckhoff & Ahn (2010), they are not repeated here.[284] Instead, GBCC-MIN and a GBCC-MAX models equivalent to the traditional BCC-I and BCC-O models that assign VRS are introduced in the following.[285] Regarding the respective DMU^o under evaluation, the former model has the multiplier form

$$\begin{aligned} & \max L^o = \sum_{h=1}^{r} \mu_h l_h^o - \varpi^o \\ \text{s.t.} \quad & \sum_{g=1}^{s} v_g k_g^o = 1 \\ & \sum_{h=1}^{r} \mu_h l_h^\rho - \sum_{g=1}^{s} v_g k_g^\rho - \varpi^o \le 0 \qquad \rho = 1,\ldots,\pi \\ & \varpi^o \in \mathbb{R} \text{ and } v_g \ge 0,\ \mu_h \ge 0 \qquad g = 1,\ldots,s \qquad h = 1,\ldots,r \end{aligned} \tag{5.4}$$

and the envelopment form

$$\begin{aligned} & \min \theta^o \\ \text{s.t.} \quad & \sum_{\rho=1}^{\pi} \lambda^\rho l_h^\rho - l_h^o \ge 0 \qquad h = 1,\ldots,r \\ & \sum_{\rho=1}^{\pi} \lambda^\rho k_g^\rho - \theta^o k_g^o \le 0 \qquad g = 1,\ldots,s \\ & \sum_{\rho=1}^{\pi} \lambda^\rho = 1 \text{ and } \lambda^\rho \ge 0 \qquad \rho = 1,\ldots,\pi \end{aligned} \tag{5.5}$$

284 For the description of GCCR-MIN and GCCR-MAX models, see Dyckhoff & Ahn (2010), pp. 1262–1264.

285 Cf. Ahn & Dyckhoff (2011), pp. 5–6.

The GBBC-MAX model has the multiplier form

$$\begin{aligned}
&\min K^o = \sum_{g=1}^{s} v_g k_g^o - \varpi^o \\
\text{s.t.}\quad &\sum_{h=1}^{r} \mu_h l_h^o = 1 \\
&\sum_{h=1}^{r} \mu_h l_h^\rho - \sum_{g=1}^{s} v_g k_g^\rho + \varpi^o \leq 0 \qquad \rho = 1,\ldots,\pi \\
&\varpi^o \in \mathbb{R} \text{ and } v_g \geq 0,\ \mu_h \geq 0 \qquad g = 1,\ldots,s \qquad h = 1,\ldots,r
\end{aligned} \tag{5.6}$$

and the envelopment form

$$\begin{aligned}
&\max \eta^o \\
\text{s.t.}\quad &\sum_{\rho=1}^{\pi} \lambda^\rho l_h^\rho - \eta^o l_h^o \geq 0 \qquad h = 1,\ldots,r \\
&\sum_{\rho=1}^{\pi} \lambda^\rho k_g^\rho - k_g^o \leq 0 \qquad g = 1,\ldots,s \\
&\sum_{\rho=1}^{\pi} \lambda^\rho = 1 \quad \text{and} \quad \lambda^\rho \geq 0 \qquad \rho = 1,\ldots,\pi
\end{aligned} \tag{5.7}$$

In the following, the GBBC-MAX model is specified, assuming linear cost as well as benefit objective functions. Inserting (5.2) into (5.6) and (5.7), the resulting model has the multiplier form

$$\begin{aligned}
&\min K^o = \sum_{\tau=1}^{m} \tilde{v}_\tau x_\tau^o + \sum_{\tau=1}^{n} \tilde{v}_{m+\tau} y_\tau^o - \varpi^o \\
\text{s.t.}\quad &\sum_{\tau=1}^{m} \tilde{\mu}_\tau x_\tau^o + \sum_{\tau=1}^{n} \tilde{\mu}_{m+\tau} y_\tau^o = 1 \\
&\sum_{\tau=1}^{m} (\tilde{\mu}_\tau - \tilde{v}_\tau) x_\tau^\rho + \sum_{\tau=1}^{n} (\tilde{\mu}_{m+\tau} - \tilde{v}_{m+\tau}) y_\tau^\rho + \varpi^o \leq 0 \qquad \rho = 1,\ldots,\pi \\
&\tilde{v}_\tau = \sum_{g=1}^{s} v_g c_{g,\tau} \text{ and } \tilde{\mu}_\tau = \sum_{h=1}^{r} \mu_h e_{h,\tau} \qquad \tau = 1,\ldots,m+n \\
&\varpi^o \in \mathbb{R} \text{ and } v_g \geq 0,\ \mu_h \geq 0 \qquad g = 1,\ldots,s \qquad h = 1,\ldots,r
\end{aligned} \tag{5.8}$$

and the envelopment form

$$
\begin{aligned}
&\max \eta^{o} \\
\text{s.t.}\ &\sum_{\tau=1}^{m} e_{h,\tau}\left(\tilde{x}_{\tau}-\eta^{o}x_{\tau}^{o}\right)+\sum_{\tau=1}^{n} e_{h,m+\tau}\left(\tilde{y}_{\tau}-\eta^{o}y_{\tau}^{o}\right)\geq 0 \qquad h=1,\ldots,r \\
&\sum_{\tau=1}^{m} c_{g,\tau}\left(\tilde{x}_{\tau}-x_{\tau}^{o}\right)+\sum_{\tau=1}^{n} c_{g,m+\tau}\left(\tilde{y}_{\tau}-y_{\tau}^{o}\right)\leq 0 \qquad g=1,\ldots,s \\
&\tilde{x}_{\tau}=\sum_{\rho=1}^{\pi}\lambda^{\rho}x_{\tau}^{\rho} \quad \tau=1,\ldots,m \quad \text{and} \quad \tilde{y}_{\tau}=\sum_{\rho=1}^{\pi}\lambda^{\rho}y_{\tau}^{\rho} \quad \tau=1,\ldots,n \\
&\sum_{\rho=1}^{\pi}\lambda^{\rho}=1 \quad \text{and} \quad \lambda^{\rho}\geq 0 \qquad \rho=1,\ldots,\pi
\end{aligned}
\tag{5.9}
$$

The generalization of traditional DEA may not be apparent in the GDEA models presented in (5.4) – (5.7), as they seem to be similar to traditional models with only a difference in the use of variables – k and l instead of x and y. However, considering the specification in (5.8) and (5.9) for the case of linear cost and benefit functions, the different impact of the GDEA models becomes clear. For instance, new multiplier variables $\tilde{\nu}_{\tau}$ and $\tilde{\mu}_{\tau}$ for the inputs and outputs are introduced in (5.8) to represent cost and benefit multipliers, respectively. A cost multiplier $\tilde{\nu}_{\tau}$ results from the sum of the products of the multiplier of the cost objective and the corresponding input as well as output multiplier for each cost objective. Analogously, a benefit multiplier $\tilde{\mu}_{\tau}$ is derived from the multipliers of the benefit objective. In the special case that every cost objective is reflected only by one input and every benefit objective by one output, both GDEA models – GCCR and GBCC – will be identical to the respective well-known traditional DEA models – CCR and BCC.[286]

Nevertheless, it is obvious that the specification and application of GDEA models represent a challenging task that can easily become a complex problem. Without methodical support, it is hardly possible for the decision makers to establish a valid GDEA model for the respective application case.[287] Therefore, it seems reasonable to provide a stepwise approach that assists the decision makers in developing an appropriate GDEA model based on the determination of relevant objectives and corresponding performance criteria required for the efficiency evaluation. The following section introduces such a procedure by utilizing the concepts of MCDM for determining, structuring, and operationalizing objectives in the GDEA context.

286 Cf. Ahn & Dyckhoff (2011), p. 7; Dyckhoff & Ahn (2010), p. 1268.
287 Cf. Ahn & Dyckhoff (2011), pp. 17–18; Dyckhoff & Ahn (2010), pp. 1269–1270.

5.2 Goal-oriented approach to determine performance criteria for evaluating relative efficiency

5.2.1 Framework

For an exemplary elucidation, the framework is presented on the basis of an explicit example – the case of a European pharmacy chain – already referred to in previous chapters. It is assumed that the central management of the pharmacy chain compares the efficiency of its stores with regard to one of their fundamental goals – strengthening of customer retention – mentioned in Subsection 3.3.3. With the help of this (fictive) example, the problem solving potential of the approach is described in the next subsection. It should be noted that the focus of this section is on procedural aspects to generate the results, not on the results themselves, particularly because they depend to a large extent on the preferences of the decision makers.

The resulting three-step procedure to determine suitable performance criteria for the evaluation of relative efficiency from a goal-oriented perspective is presented in the next subsection. By using MCDM concepts, the procedure suggests firstly to structure the relevant organizational goals, then to determine appropriate performance criteria which allow an operationalization of the goals, and finally to quantitatively describe the linkage between the goals and criteria through linear cost and benefit functions.

5.2.2 Three-step procedure

Strengthening of customer retention can be considered as one of the goals of a pharmacy chain's BSC.[288] To evaluate the efficiency with which such a fundamental goal is achieved, the following three steps should be carried out first: development of a system of appropriate lower-level goals, derivation of suitable performance criteria, and construction of functional relationships between the determined goals and respective criteria.

5.2.2.1 Step 1: Development of a system of objectives

In a first step, a simple but complete and at the same time redundance-free system of lower-level goals has to be derived. The term *system* indicates that it is recommended to structure the lower-level goals – hereinafter called objectives.

288 Cf. Kaplan & Norton (2003), no page; Shutt (2003), pp. 43–44.

In the context of GDEA, it is meaningful to divide them into cost objectives to be minimized and benefit objectives to be maximized.

Referring to the example of pharmacy stores, the group of relevant cost objectives comprises the *cost of service* for the customers, the customers' *waiting time in the store* and their *waiting time for drugs not in stock*, as shown in Figure 5.1. Concerning the relevant benefit objectives, it is distinguished between the *customer advisory service concerning drugs* and the *customer advisory service concerning over-the-counter sales*, provided by the service personnel of the pharmacy stores. It is emphasized once more that this is just an exemplary system of objectives to measure the functional or rather partial efficiency of strengthening the customer retention.

Fig. 5.1: Splitting the fundamental goal of strengthening the customer retention in lower-level goals[289]

Goal level	Fundamental goal: strengthening of customer retention				
	lower-level goals (= objectives) to be minimized			lower-level goals (= objectives) to be maximized	
	k_1: cost of service	k_2: waiting time in the store	k_3: waiting time for drugs not in stock	l_1: customer advisory service concerning drugs	l_2: customer advisory service concerning over-the-counter sales

The MCDM literature provides some basic concepts to facilitate the development of such a system.[290] Especially, it is recommended to build a hierarchy of objectives, following a set of rules, e.g., breaking down a goal or objective into its constituent parts helps to avoid redundancy between the lower-level objectives and to ensure that all relevant aspects are covered. Although the results will always be influenced by individual preferences, the structured procedure to derive a system of objectives helps third parties to understand and evaluate these results.

5.2.2.2 Step 2: Derivation of suitable performance criteria

The second step refers to the need of determining performance criteria which allow the quantification of the achieved values k_g and l_h of the identified objectives. In the MCDM literature, these performance criteria are often called attributes,

289 Cf. Afsharian et al. (2016), p. 1896.

290 Cf. Keeney & Raiffa (1993), Chapter 2; Eisenführ et al. (2010), Chapter 3.

whereby natural, proxy, and artificial ones are distinguished.[291] Table 5.1 describes these kinds of attributes and their occurrence in the example.

Tab. 5.1: Characterization of attributes as performance criteria to measure the achievement of objectives[292]

Kind of attribute	Description	Example	Remarks
Natural attribute	Arises directly from the objective under consideration and provides an unambiguous measurement	Objective: cost of service Attributes: *cost of service personnel* and *cost of expired drugs*	The prerequisite is that these attributes are really equivalent to the cost of service
Proxy attribute	An indicator or a means for the achievement of an objective	Objective: waiting time in the store Attribute: *number of complaints about waiting time in the store* Objective: waiting time for drugs not in stock Attribute: *number of subsequent drug deliveries* Objective: customer advisory service concerning over-the-counter sales Attribute: *turnover of over-the-counter sales*	Data availability may be the reason why not the waiting time of the customers in the store, waiting time of the customers for drugs, and outcome of customer advisory service concerning over-the-counter sales themselves are measured (as natural attributes), but the respective complaints, subsequent drug deliveries, and the turnover of over-the-counter sales
Artificial attribute	A constructed combination of criteria relevant for the objective	Objective: customer advisory service concerning drugs Attributes: *telephone hotline hours for health counselling services* and *number of health counselling services in the store*	Artificial attributes can be seen as a combination of several proxy attributes

291 Cf. Eisenführ et al. (2010), pp. 72–73.
292 Afsharian et al. (2016), p. 1897.

Most attributes in the example can be classified as inputs or outputs of the business processes in a pharmacy store, as shown in Figure 5.2. However, such an unambiguous characterization is not possible for the *number of health counselling services in the store*. As already mentioned, this attribute refers to a service process which typically encompasses at the same time input and output properties. The same would apply to *telephone hotline hours for health counselling services*, if not the stand-by time would be meant by this attribute but the actual counselling time.

Fig. 5.2: Classification of suitable performance criteria for the identified objectives[293]

Goal level	Fundamental goal: strengthening of customer retention				
	lower-level goals (= objectives) to be minimized			lower-level goals (= objectives) to be maximized	
	k_1: cost of service	k_2: waiting time in the store	k_3: waiting time for drugs not in stock	l_1: customer advisory service concerning drugs	l_2: customer advisory service concerning over-the-counter sales

Performance criteria level	CSP: cost of service personnel	CED: cost of expired drugs	HHC: telephone hotline hours for health counselling services	HCS: number of health counselling services in the store	CWT: number of complaints about waiting time in the store	SDD: number of subsequent drug deliveries	TOTC: turnover of over-the-counter sales
	inputs				outputs		

In any case, the classification of the attributes as inputs or outputs is not crucial when measuring goal-oriented efficiency. The reason is that the MCDM approach to DEA efficiency evaluation abstains from splitting up a production (or service) process into an input, throughput, and output phase. In contrast, the traditional DEA is based on such a three-phase point of view. However, this is not a methodical advantage as long as the throughput phase is "regarded as a black box"[294].

293 Cf. Afsharian et al. (2016), p. 1896.
294 Rousseau & Rousseau (1997), p. 47.

5.2.2.3 Step 3: Construction of cost and benefit functions

In the third step, the qualitative relation between objectives and attributes has to be converted into value functions, i.e. cost functions and benefit functions, which describe the quantitative linkage between them. In the best case, objectives are described by natural attributes, which means that the respective value functions are obvious. For instance, it will be generally agreed that the relation between the objective *cost of service* (k_1) and its attributes *cost of service personnel* (CSP) and *cost of expired drugs* (CED) should be quantitatively described by the linear cost function k_1 = CSP + CED.

In the case of a proxy attribute, the respective value function is subject to an individual estimation of the quantitative linkage between the objective and its attribute/s. For example, the objective *waiting time for drugs not in stock* (k_3) is modeled by a cost linear function k_3 = SDD of the attribute *number of subsequent drug deliveries* (SDD). The latter can be characterized as an output to be minimized. Unlike in traditional DEA, this undesirable performance criterion does not cause any methodical problems in the goal-oriented DEA approach. Analogously, the objective *waiting time in the store* (k_2) is reflected by a cost linear function k_2 = CWT and the benefit objective *customer advisory service concerning over-the-counter sales* (l_2) is modeled by the respective linear function l_2 = TOTC.

Artificial attributes are a special case in the sense that they measure an objective per definitionem by a function of at least two different performance criteria. In the example, the *customer advisory service concerning drugs* (l_1) is quantified by a linear benefit function of the *telephone hotline hours for health counselling* services (HHC_{norm}) and the *number of health counselling services in the store* (HCS_{norm}); thereby, normalized values of these two attributes are suggested to reduce scale size effects on the results: $l_1 = HHC_{norm} + HCS_{norm}$.

Any normalization should be based on reasonable attribute ranges. For the *telephone hotline hours for health counselling services*, e.g., a range of [0, 720] seems appropriate with respect to a time horizon of one month with 30 days and a single hotline which can be operated between zero and (30 days • 24 hours per day =) 720 hours. If such natural limits do not exist, a realistic attribute range has to be estimated. Alternatively, the range resulting from the given data set may be considered. However, this will cause range size effects when additional DMUs with extreme attribute values are subsequently included in the analysis.

Figure 5.3 summarizes the three-steps of the delineated procedure for the determination of performance criteria from a goal-oriented perspective. The construction of the exemplarily described cost and benefit functions – in Figure 5.3 labeled as GDEA level – is a prerequisite to apply GDEA. In the next subsection, an exemplary specification of a GDEA model suitable for the presented example is illustrated.

Fig. 5.3: Determination of performance criteria for DEA from a goal-oriented perspective: the example of strengthening of customer retention[295]

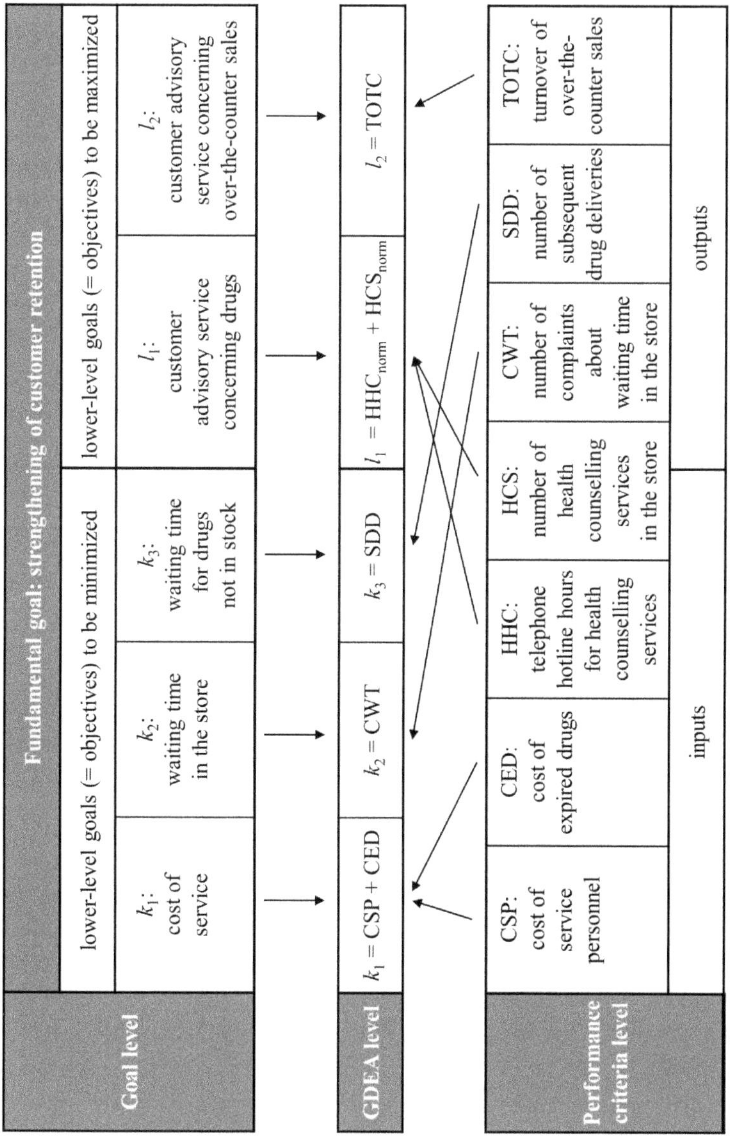

295 Afsharian et al. (2016), p. 1896.

5.2.3 Illustrative example

Based on the goal-oriented procedure for the determination of performance criteria presented in previous subsection and summarized in Figure 5.3, the theoretical concept of GDEA described in Subsection 5.1.3 is tailored to the case of evaluating the pharmacy stores' functional efficiency.

As the focus of the evaluation is on the strengthening of customer retention, the improvement of benefit objectives, i.e. the lower-level objectives to be maximized, is emphasized. Additionally, it is assumed that the pharmacy stores – as the π DMUs under consideration ($\rho = 1,\ldots,\pi$)– operate under variable returns to scale. For this scenario, a GBBC-MAX model seems to be appropriate to evaluate the goal-oriented efficiency. The respective GBCC-MAX model in the multiplier form reads:

$$
\begin{aligned}
&\min K^o = v_1 k_1^o + v_2 k_2^o + v_3 k_3^o - \varpi^o \\
\text{s.t.}\quad & \mu_1 l_1^o + \mu_2 l_2^o = 1 \\
& \mu_1 l_1^\rho + \mu_2 l_2^\rho - v_1 k_1^\rho - v_2 k_2^\rho - v_3 k_3^\rho + \varpi^o \le 0 \qquad \rho = 1,\ldots,\pi \\
& \varpi^o \in \mathbb{R} \text{ and } v_g \ge 0,\ \mu_h \ge 0 \qquad g = 1,\ldots,3 \qquad h = 1,\ldots,2
\end{aligned}
\tag{5.10}
$$

or, with respect to (5.2),

$$
\begin{aligned}
&\min K^o = v_1\left(\mathrm{CSP}^o + \mathrm{CED}^o\right) + v_2\mathrm{CWT}^o + v_3\mathrm{SDD}^o - \varpi^o \\
\text{s.t.}\quad & \mu_1\left(\mathrm{HHC}^o_{\mathrm{norm}} + \mathrm{HCS}^o_{\mathrm{norm}}\right) + \mu_2\mathrm{TOTC}^o = 1 \\
& \mu_1\left(\mathrm{HHC}^\rho_{\mathrm{norm}} + \mathrm{HCS}^\rho_{\mathrm{norm}}\right) + \mu_2\mathrm{TOTC}^\rho \\
& \qquad - v_1\left(\mathrm{CSP}^\rho + \mathrm{CED}^\rho\right) - v_2\mathrm{CWT}^\rho - v_3\mathrm{SDD}^\rho + \varpi^o \le 0 \\
& \rho = 1,\ldots,\pi \\
& \varpi^o \in \mathbb{R} \text{ and } v_g \ge 0,\ \mu_h \ge 0 \qquad g = 1,\ldots,3 \qquad h = 1,\ldots,2
\end{aligned}
\tag{5.11}
$$

While in (5.2), it is distinguished between inputs and outputs, this distinction is not necessary to formulate (5.11). Especially, the $\mathrm{HHC}_{\mathrm{norm}}$ criterion (normalized telephone hotline hours for health counselling services) can be taken into account in the model without problems, although its characterization as input or output is not unequivocally possible. The envelopment form of the GBCC-MAX model reads:

$$\begin{aligned}
&\max \eta^o \\
\text{s.t.}\quad & \sum_{\rho=1}^{\pi} \lambda^\rho l_1^\rho - \eta^o l_1^o \geq 0 \\
& \sum_{\rho=1}^{\pi} \lambda^\rho l_2^\rho - \eta^o l_2^o \geq 0 \\
& \sum_{\rho=1}^{\pi} \lambda^\rho k_1^\rho - k_1^o \leq 0 \\
& \sum_{\rho=1}^{\pi} \lambda^\rho k_2^\rho - k_2^o \leq 0 \\
& \sum_{\rho=1}^{\pi} \lambda^\rho k_3^\rho - k_3^o \leq 0 \\
& \sum_{\rho=1}^{\pi} \lambda^\rho = 1 \quad \text{and} \quad \lambda^\rho \geq 0 \qquad \rho = 1,\ldots,\pi
\end{aligned} \tag{5.12}$$

or, with respect to (5.2),

$$\begin{aligned}
&\max \eta^o \\
\text{s.t.}\quad & \sum_{\rho=1}^{\pi} \lambda^\rho \left(\mathrm{HHC}_{\mathrm{norm}}^\rho + \mathrm{HCS}_{\mathrm{norm}}^\rho\right) - \eta^o \left(\mathrm{HHC}_{\mathrm{norm}}^o + \mathrm{HCS}_{\mathrm{norm}}^o\right) \geq 0 \\
& \sum_{\rho=1}^{\pi} \lambda^\rho \mathrm{TOTC}^\rho - \eta^o \mathrm{TOTC}^o \geq 0 \\
& \sum_{\rho=1}^{\pi} \lambda^\rho \left(\mathrm{CSP}^\rho + \mathrm{CED}^\rho\right) - \left(\mathrm{CSP}^o + \mathrm{CED}^o\right) \leq 0 \\
& \sum_{\rho=1}^{\pi} \lambda^\rho \mathrm{CWT}^\rho - \mathrm{CWT}^o \leq 0 \\
& \sum_{\rho=1}^{\pi} \lambda^\rho \mathrm{SDD}^\rho - \mathrm{SDD}^o \leq 0 \\
& \sum_{\rho=1}^{\pi} \lambda^\rho = 1 \quad \text{and} \quad \lambda^\rho \geq 0 \qquad \rho = 1,\ldots,\pi
\end{aligned} \tag{5.13}$$

Translating the considerations summarized in Figure 5.3 into the formula, the models (5.10) to (5.13) provide an efficiency score for each DMU which can be referred to as goal-oriented efficiency concerning the strengthening of customer retention. The example shows how GDEA broadens the traditional DEA approach by combining its focus on performance criteria with the goal-oriented focus of MCDM. The resulting managerial implications are discussed in the following.

5.3 Managerial implications

The determination of performance criteria by means of an explicit reference to organizational goals provides the basis for a thorough performance evaluation. The goal-oriented perspective ensures that criteria used for the evaluation are selected based upon their ability to reflect the organizational goals. Thus, the subsequent evaluation is able to show proper results, which can provide meaningful information concerning the success of the organization in aligning its performance in accordance with its goals.

Furthermore, the proposed step-by-step procedure creates transparency concerning the determination of relevant performance objectives and respective criteria. This enhances the comprehensibility of the results of the corresponding performance evaluation, especially for external third parties. In addition, the exemplary illustration of the approach reveals the cause of the outlined problems in the performance criteria determination: it is decisive that DEA is just a special case of GDEA. While GDEA requires to explicitly specify cost and benefit functions, this is only implicitly done in traditional DEA, with the very strict assumption of linear functions in the form of $k_g = x_g$ and $l_h = y_h$ ($g = 1,\ldots,s$ and $h = 1,\ldots,r$). Refraining from this assumption, the GDEA approach can contribute to coping with the three outlined problems in the performance criteria determination:

- *The problem of performance criteria selection*
  The relevance of performance criteria depends on the objectives to be pursued by the DMUs. Although this may seem obvious, many DEA applications give rise to suspect that data are not preliminary chosen with regard to their appropriateness, but due to their availability. In contrast, a sound DEA application requires first to systematically derive the objectives to be taken into account. This represents a prerequisite for the next step of selecting suitable performance criteria. The approach proposed in the previous subsection refers exactly to these steps. Therefore, this problem does not occur in the context of GDEA.
- *The problem of dual-role performance criteria*
  As emphasized in the context of (5.11), a GDEA application does not depend on the possibility to characterize performance criteria either as inputs or as outputs. From an MCDM point of view, one could entirely abstain from distinguishing between inputs and outputs. However, such a differentiation makes

sense from the perspective of production theory regarding analyses beyond GDEA efficiency.
While a natural attribute arises unambiguously from the objective under consideration, proxy, and artificial attributes leave room for the decision makers how to choose and construct them, respectively. As a result, e.g., the same attribute may appear twice in the performance criteria set. A well-known example recognized in DEA literature refers to bank deposits and the ongoing discussion, if they should be regarded as inputs or outputs.[296] From the GDEA point of view, the answer to this question is that it depends on their power to reflect the actual performance objective/s.

- *The problem of anti-isotonic performance criteria*
  In analogy to dual-role performance criteria, the problem of undesirable criteria occurs only if proxy or artificial attributes are used. Taking the example of the *waiting time in the store* as objective k to be minimized, the natural attribute to measure this objective would be the actual waiting time itself. If corresponding data is not available, the *number of complaints about waiting time in the store* (CWT) could be considered as an appropriate proxy attribute. In DEA, the problem would arise that this attribute is an output to be minimized. This problem can be solved by using the cost function $k = \text{CWT}$ in GDEA.

5.4 Conclusion and discussion

This chapter has addressed the managerial and procedural problem of determining suitable performance criteria for the evaluation of relative efficiency of DMUs. As the DEA methodology does not include any methodical procedure for the determination of suitable performance criteria and due to the focus of DEA efficiency evaluation on the transformation process of inputs into outputs, assuming them to be isotonic, three kinds of problems in criteria determination have been described. The explicit incorporation of organizational goals within the evaluation process and their connection with corresponding performance criteria has been detected as a solution to the outlined problems, requiring a rethinking of DEA efficiency evaluation. Therefore, the idea of Dyckhoff & Ahn (2010) to extend the DEA methodology by GDEA, which adjusts traditional DEA models to the evaluation of goals, has been presented.

296 Cf. Fethi & Pasiouras (2010), pp. 191–192.

Based on a brief review of GDEA methodology, the importance of a methodical support for the decision makers that facilitates the implementation of GDEA has been emphasized. Thus, a stepwise approach that assists the decision makers in developing an appropriate GDEA model based on the determination of suitable and measurable performance criteria from goal-oriented perspective have been introduced. In doing so, the basic concepts of MCDM for determining, structuring, and operationalizing objectives have been utilized in the GDEA context. The systematic selection of performance criteria with relevance to the goals enables a thorough performance evaluation and confirms the ability of GDEA to handle the outlined problems in the performance criteria determination.

From a managerial point of view, the goal-oriented approach to determine performance criteria for the evaluation of relative efficiency presented within this chapter is aimed to meet the requirements on model design provided by Little (2004). Its strengths and limitations referring to these requirements are presented in Table 5.2. In detail, the presented step-by-step procedure facilitates the determination of relevant objectives and corresponding performance criteria required for the efficiency evaluation. However, the specification of functional relationships between determined objectives and respective criteria can be very challenging for the decision makers, especially if the objective/s cannot be described by natural attribute/s. In such cases, the decision makers have to properly estimate a quantitative linkage between objectives and criteria.

By utilizing the basic MCDM concepts for determining, structuring, and operationalizing objectives, the approach allows a robust and replicable determination of performance objectives and criteria. However, the resulted system of objectives and corresponding value functions will always be influenced by individual preferences of the decision makers, on the one side, and be sensitive do data availability with regard to the criteria used as attributes for the quantification of objectives, on the other.

Obviously, the awareness and use of basic MCDM concepts by the decision makers is an essential prerequisite for the success of the proposed approach. It allows the decision makers to easily control the procedure through the incorporation of organizational goals as decisive performance objectives and all kinds of performance criteria, independently whether they are isotonic, anti-isotonic and/or have a dual form. Thus, without the methodical support of MCDM, it would be difficult for the decision makers to determine a simple, complete, and redundance-free system of objectives and respective criteria.

Tab. 5.2: Strengths and limitations of the presented goal-oriented approach to determine performance criteria for evaluating relative efficiency

Managerial requirements on model design[297]	**Strengths**	**Limitations**
Simple	Step-by-step determination of relevant objectives and their connection with measurable performance criteria	Specification of functional relationships between determined performance objectives and respective criteria
Robust	Replicable procedure based on basic MCDM concepts	Sensitive to data availability and individual preferences of the decision makers
Easy to control	Incorporation of organizational goals as decisive performance objectives and all kinds of performance criteria, independently whether they are isotonic, anti-isotonic and/or have a dual form	Based on the assumption that MCDM concepts for the determination and structure of the objectives are known and used by the decision makers
Adaptive	Different units of measurement are acceptable	
Complete on important issues	Goal-oriented perspective in the determination of relevant performance objectives and respective criteria	Subjectivity concerning the selection of performance objectives and respective criteria
Easy to communicate with	Transparency and comprehensibility of the determination process for external third parties	

As the determination of performance objectives and criteria is aimed for the evaluation of goal-oriented efficiency by means of an appropriate GDEA model, the presented approach keeps the beneficial feature of the DEA methodology to handle different units of measurement, being, in this way, adaptive to new information. Moreover, the incorporation of a goal-oriented perspective in the determination of relevant performance objectives and respective criteria allows the approach to be complete in important issues. Although, subjective judgments

297 Cf. Little (2004), pp. 1843–1844.

of the decision makers concerning the selection of performance objectives and corresponding criteria cannot be avoided, they will become more apparent than in the case of traditional DEA. However, if the fundamental goals of multiple stakeholders of an organization have to be simultaneously included in the efficiency evaluation, an extension of the presented approach might be useful. This could be based on, e.g., the Decision-oriented Performance Measurement framework of Ahn & Le (2015) – a hierarchical four-step process based on rationality concepts of decision-making – which ensures that the performance of an organization is linked to stakeholder values.

Finally, the main benefit of the proposed approach is its step-by-step created transparency of the criteria selection process. The insights provided by the approach allow external third parties to better understand the underlying assumptions and, therefore, to assess and/or accept the evaluation results. This is an important step forward in advancing the use of GDEA in practice.

6 Conclusions

6.1 Summary and conclusions

Performance evaluation has become part of the business' and research's daily life, leading to numerous possibilities to evaluate performance. While practitioners usually refer to accounting based measurement systems, scientists rather focus on new approaches on performance evaluation. The disadvantage of – in organizations widely used – accounting based measurement systems is their inability to consider multiple performance measures defined in different units of measurement for the evaluation of organizational performance. Taking multiple measures in consideration, however, is critical for organizations to evaluate themselves in today's multi-facet business environment. By looking into the vast amount of approaches available for performance evaluation, one sticks out as able to meet the challenge in aggregating all relevant performance measures – independent of their units of measurement – into a single comprehensive efficiency score: the data envelopment analysis (DEA).

After being introduced into the scientific community in 1978, DEA has been researched and enhanced a lot from a scientific perspective. However, the importance of DEA within international research cannot be seen as an indicator for its acceptance or use in business practice. Practitioners often do not realize the usefulness of applying DEA. Further, the need to familiarize themselves with the mathematical background of DEA inhibits the attractiveness of DEA for practitioners.

In order to close the gap between DEA research and its recognition and utilization in business practice, the practitioners' requirements should be given more attention within the scientific research. Therefore, the aim of this dissertation has been to propose new methodological developments of DEA that satisfy the demands of business practice and provide a multidimensional point of view on the evaluation of organizational performance. This undertaking has been backed by experience and data gained within a research project between a European pharmacy chain and the Institute of Management Control and Business Accounting. This collaboration has given interesting insights into the requirements of business practice. Further, the modified data partially obtained within this project has formed the basis for a consistent but fictive example used throughout the dissertation in order to illustrate the introduced approaches.

To begin with, an integrated concept of performance management has been defined. On the one hand, this has been aimed for a clear differentiation between the roles of performance measurement, performance evaluation, and performance management in general. On the other hand, performance evaluation has been detected as the element of performance management, in which the potential of DEA for managerial decision making could be embedded. By using the procedure of Golany & Roll (1989) as a general guideline for modelling DEA applications, DEA has been incorporated into the process of performance evaluation. This link has resulted in a comprehensive and systematic step-by-step framework for performance evaluation based on DEA. While analyzing the framework for the possible potential of methodological developments of DEA three issues entailing managerial problems could be revealed: the limited focus of DEA on the evaluation of relative efficiency as only one dimension of performance, the possibility of neglecting some of the performance criteria within the evaluation, and the difficulty in determining suitable performance criteria. A solution to each of these problems has been derived from a multidimensional point of view on the evaluation of organizational performance.

The first problem has been addressed by extending the DEA methodology for the evaluation of effectiveness as a further meaningful performance dimension. According to Weber's (1978) treatise of purpose-rational behavior, effectiveness refers exclusively to the original goals of economic activities. Such goals are operationalized within most of the companies through the use of common performance management systems. In these cases, relevant information concerning indicators of goal and respective targets used for the quantification of the goals is already available. Referring to the existence of information about predetermined targets, a new DEA approach for the evaluation of relative effectiveness has been developed. Based on the well-known traffic light system, the approach classifies decision making units (DMUs) into subsets according to their ability to meet the predetermined targets and provides a complete ranking within each subset. This allows a simple visualization of the results that enhances the comprehensibility of the approach and supports managerial decision making, e.g. inducing plausible managerial incentives or indicating whether the predetermined targets are appropriate. Moreover, as the achievement of goals beyond the predetermined targets is usually welcomed, the concept of super-effectiveness has been introduced, implementing the requirement for a balanced fulfillment of the targets.

The problem of a possible neglect of crucial performance criteria within the DEA evaluation has been addressed by introducing a measure of balance – as a further dimension of performance. Based on a typical data set for a DEA application, the balance score evaluates the extent to which a DMU avoids concentration on only some of the crucial performance criteria and goals. This additional information about a DMU's performance structure not only permits the evaluation of the comparability of the DMUs under analysis, but also questions the role of efficient but unbalanced DMUs. In the latter case, a modified CCR-O model has been proposed that avoids the use of such DMUs as benchmarks for other DMUs. Thus, an iterative process of balance and efficiency analysis leads to adequate efficiency scores and supports the unbalanced DMUs on their way to get balanced. The value-added potential of the idea of balance as an additional performance measure has been recognized by other scientists, encouraging them to develop alternative models for the evaluation of relative balance.

The last problem concerning the determination of suitable performance criteria has been addressed by introducing a stepwise approach for the selection of appropriate criteria from goal-oriented perspective of multi-criteria decision making (MCDM). This perspective ensures that criteria used for the evaluation are selected based upon their ability to adequately reflect the organizational goals. A characterization of these criteria as inputs or outputs is not crucial, allowing the incorporation of anti-isotonic and/or dual-role criteria in the evaluation. Consequently, DEA has to be seen in a broader context, following the recommendation to expand the traditional DEA by a generalized DEA (GDEA). GDEA combines the DEA focus on performance criteria with the goal-oriented focus of MCDM by adjusting traditional DEA models to the evaluation of goals instead of inputs and outputs. The determination of performance criteria according to the goal-oriented perspective, which is part of GDEA, provides the basis for a thorough performance evaluation and creates transparency of the criteria selection process. The insights provided by the approach allow external third parties to better understand the underlying – mostly implicitly – subjective assumptions and, therefore, to assess and/or accept the results of the evaluation. This is an important step forward in advancing the use of GDEA and DEA, respectively, in practice.

Finally, the presented approaches extend the focus of a performance evaluation based on DEA, which is originally placed on efficiency, to effectiveness and balance – as two additional dimensions of performance. Such multidimensional point of view provides different insights into organizational performance, which by complementing each other have several implications for managerial decision making. The discussed managerial implications of each approach confirm the relevance of DEA for the management of organizational performance not only as a practicable tool for performance evaluation, but also as a source of important information with impact on performance control. The reflection of each approach with regard to managerial requirements on model design, proposed by

Little (2004), has implied some limitations of the approaches that might be used as starting points for further research.

6.2 Suggestions for further research

Initial recommendations for future research have been implied in the individual chapters of the dissertation. In Chapter 3, the limited assumption of one predetermined target vector as a benchmark for the evaluation of effectiveness has been mentioned. The multiple stakeholder of an organization may, however, have different expectations on appropriate target levels for goal attainment. Further research should, therefore, investigate the possibility of evaluating effectiveness on the basis of multiple predetermined target vectors.

Due to the new idea of balance as an additional performance measure, introduced in Chapter 4, there is significant potential for further developments, especially concerning the selection of performance criteria in which a degree of balance is preferable from a managerial point of view. In Chapter 4, the balance of the output mix is prioritized, referring to outputs as appropriate indicators of the fundamental goals of an organization. However, as described in Chapter 5, relevant organizational goals could be operationalized by inputs and outputs simultaneously. This requires a modification of the presented approach in order to evaluate the balance of an input-output mix. Moreover, corresponding to the explicit reference of GDEA to organizational goals as decisive performance objectives, the possibility of extending the focus of balance evaluation to the examination of goals instead of inputs and/or outputs should be considered in future investigations.

Broadening the DEA methodology by GDEA extensively explicated by Dyckhoff & Ahn (2010) and shortly described in Chapter 5 raises additional interesting research issues. From the scientific perspective, the development of unoriented GDEA models may be necessary in cases where the emphasis is placed on a conjointly attainment of goals to be maximized and goals to be minimized. Also, supplemental analyses, which are typical of traditional DEA applications, e.g., correlation studies, should be reflected in the light of GDEA. From the practical perspective, it may be worth considering multi-period evaluations where short-term and long-term goals have to be simultaneously taken into account.

The practicability of the approaches developed within this dissertation has been illustrated by an example partially based on real-world data. However, the implementation of the whole framework – including the presented approaches – by practitioners has not yet been done. This should be supported by the development of a user friendly software interface that not only guides the user through each step of the framework, but also initially requests all relevant information for performance evaluation from a multidimensional point of view. In doing so, the software could simultaneously evaluate the different performance dimensions in order to deliver quick results. In this context, the common prerequisite of the approaches that organizational goals are linked to pertinent routine performance information in terms of measurable performance criteria, which constitute the basis for the respective evaluations, could be made fruitful. The practitioners should only be able to characterize their organizational goals into different goal levels – fundamental goals for effectiveness evaluation and lower-level goals for efficiency and/or balance evaluation – and provide respective goal indicators.

This dissertation has selectively addressed some shortcomings of DEA. The elimination or prevention of further DEA pitfalls would be beneficial for promotion of the use of DEA in business practice. Additionally, it should be investigated whether there are other supplementary dimensions of performance that should be integrated into the framework for performance evaluation based on DEA. For instance, the adaptability of an organization to its environment has already attracted scientific interest as an additional dimension of performance.[298] Taking adaptability into account could provide interesting information about the ability of an organization to adjust its strategic and operative management to the changes of today's dynamic environment in order to perform successfully in the long-term. In general, the main challenge of research in DEA remains in the development of practicable approaches that successfully overcome the gap between the requirements on a scientific contribution and the demands of business practice.

298 Cf. Clark (2000), pp. 6–7; Tuominen et al. (2004), p. 496.

References

Adler, R.W. (2011). Performance management and organizational strategy: how to design systems that meet the needs of confrontation strategy firms. *The British Accounting Review,* 43(4), 251–263.

Adler, N., Friedman, L., & Sinuany-Stern, Z. (2002). Review of ranking methods in data envelopment analysis context. *European Journal of Operational Research,* 140(2), 249–265.

Afsharian, M., Ahn, H., & Neumann, L. (2016). Generalized DEA: an approach for supporting input/output factor determination in DEA. *Benchmarking: An International Journal,* 23(7), 1892–1909.

Afsharian, M., Ahn, H., & Vazquez Novoa, N. (2014). Behavioral performance evaluation. In J. Wang (Ed.), *Encyclopedia of Business Analytics and Optimization* (pp. 251–264). Hershey/PA: Business Science Reference, IGI Global.

Aguinis, H. (2009). An expanded view of performance management. In J. W. Smither & M. London (Eds.), *Performance management: putting research into practice* (pp. 1–43). San Francisco: Wiley.

Aguinis, H. (2013). *Performance management.* Upper Saddle River, NJ: Pearson Prentice Hall.

Ahn, H. (1997). *Optimierung von Produktionsentwicklungsprozessen: Entscheidungsunterstützung bei der Umsetzung des Simultaneous Engineering.* Wiesbaden: Deutscher Universitäts-Verlag.

Ahn, H. (2003). Effektivitäts- und Effizienzsicherung – Controlling-Konzept und Balanced Scorecard. Frankfurt am Main: Peter Lang.

Ahn, H., Blattner, J., & Neumann, L. (2014). Unterstützung der DEA-Modellauswahl – Ein pragmatisches Konzept und MS Excel Tool. *Wirtschaftswissenschaftliches Studium,* 43(2), 111–115.

Ahn, H., & Dyckhoff, H. (2004). Zum Kern des Controllings – Von der Rationalitätssicherung zur Effektivitäts- und Effizienzsicherung. In E. Scherm, & G. Pietsch (Eds.), *Controlling – Theorien und Konzeptionen* (pp. 501–525). München: Vahlen.

Ahn, H., & Dyckhoff, H. (2011). Methodology of DEA-based performance analysis – foundations and the example of research performance. Last access: http://ssrn.com/abstract=1992367.

Ahn, H., & Le, M.H. (2014). An insight into the specification of the input-output set for DEA-based bank efficiency measurement. *Management Review Quarterly,* 64(1), 3–37.

Ahn, H., & Le, M.H. (2015). DEA efficiency of German savings banks: evidence from a goal-oriented perspective. *Journal of Business Economics,* 85(9), 953–975.

Ahn, H., & Neumann, L. (2014a). Measuring effectiveness – a DEA approach under predetermined targets. *International Journal of Business Analytics,* 1(1), 16–28.

Ahn, H., & Neumann, L. (2014b). Measuring the efficiency of pharmacy stores with respect to customer retention. In D.V. Vasilenko, G.L. Bagiev, W. Fritz, O.U. Yuldasheva, & J.N. Solovjova (Eds.), *Marketing Competencies and Business Performance: Proceedings of the International Research Symposium* (pp. 46–51). May 2014, St. Petersburg, Russia.

Ahn, H., Neumann, L., & Vazquez Novoa, N. (2012a). Measuring the relative balance of DMUs. *European Journal of Operational Research,* 221(2), 417–423.

Ahn, H., Neumann, L., & Vazquez Novoa, N. (2012b). Relative balance as complementary measure to relative efficiency. In R.D. Banker, A. Emrouznejad, A.L.M. Lopes, & M.R. de Almeida (Eds.), *Data Envelopment Analysis – Theory and Applications: Proceedings of the 10th International Conference on DEA* (pp. 265–271). August 2012, Natal, Brazil.

Ali, A.I. (1994). Computational aspects of DEA. In A. Charnes, W.W. Cooper, A.Y. Lewin, & L.M. Seiford (Eds.), *Data envelopment analysis: theory, methodology and applications* (pp. 63–88). Dordrecht: Kluwer Academic Publishers.

Alirezaee, M.R., & Afsharian, M. (2007). Model improvement for computational difficulties of DEA technique in the presence of special DMUs. *Applied Mathematics and Computation,* 186(2), 1600–1611.

Allen, R., Athanassopoulos, A., Dyson, R.G., & Thanassoulis, E. (1997). Weights restrictions and value judgments in data envelopment analysis: evolution, development and future directions. *Annals of Operations Research,* 73(0), 13–34.

Allen, R., & Thanassoulis, E. (2004). Improving envelopment in data envelopment analysis. *European Journal of Operational Research*, 154(2), 363–379.

Amaratunga, D., & Baldry, D. (2002). Moving from performance measurement to performance management. *Facilities,* 20(5/6), 217–223.

Amiri, M., Zandieh, M., Vahdani, B., Soltani, R., & Roshanaei, V. (2010). An integrated eigenvector-DEA-TOPSIS methodology for portfolio risk evaluation in the FOREX spot market. *Expert Systems with Applications,* 37(1), 509–516.

André, F.J., Herrero, I., & Riesgo, L. (2010). A modified DEA model to estimate the importance of objectives with an application to agricultural economics. *Omega,* 38(5), 371–382.

Asmild, M., Paradi, J.C., Reese, D.N., & Tam, F. (2007). Measuring overall efficiency and effectiveness using DEA. *European Journal of Operational Research,* 178(1), 305–321.

Athanassopoulos, A. (1995). Goal programming & data envelopment analysis (GoDEA) for target-based multi-level planning: allocating central grants to the Greek local authorities. *European Journal of Operational Research,* 87(3), 535–550.

Athanassopoulos, A. (1998). Decision support for target-based resource allocation of public services in multiunit and multilevel systems. *Management Science,* 44(2), 173–187.

Au, C.F. (1996). Rethinking organizational effectiveness: theoretical and methodological issues in the study of organizational effectiveness for social welfare organizations. *Administration in Social Work,* 20(4), 1–21.

Banker, R.D., Cooper, W.W., Seiford, L.M., & Zhu, J. (2004). Returns to scale in DEA. In W.W. Cooper, L.M. Seiford, & J. Zhu (Eds.), *Handbook on Data Envelopment Analysis* (pp. 41–73). Boston et al.: Kluwer Academic Publishers.

Banker, R.D., Porter, G., & Srinivasan, D. (2000). An empirical investigation of an incentive plan that includes nonfinancial performance measures. *The Accounting Review,* 75(1), 65–92.

Barnes, D., & Radnor, Z. (2008). Performance measurement and performance management: the operations management perspective. In R. Thorpe, & J. Holloway (Eds.), *Performance Management – multidisciplinary Perspectives* (pp. 89–106). London: Palgrave Macmillan.

Bentes, A.V., Carneiro, J., da Silva, J.F., & Kimura, H. (2012). Multidimensional assessment of organizational performance: integrating BSC and AHP. *Journal of Business Research,* 65(12), 1790–1799.

Bhargava, M., Dubelaar, C., & Ramaswami, S. (1994). Reconciling diverse measures of performance: a conceptual framework and test of a methodology. *Journal of Business Research, 31*(2), 235–246.

Bititci, U.S., Carrie, A.S., & McDevitt, L. (1997). Integrated performance measurement systems: an audit and development guide. *The TQM Magazine,* 9(1), 46–53.

Bititci, U.S., Garengo, P, Dörfler, V., & Nudurupati, S. (2012). Performance measurement: challenges for tomorrow. *International Journal of Management Reviews,* 14(3), 305–327.

Bititci, U.S., Turner, T., & Begemann, C. (2000). Dynamics of performance measurement systems. *International Journal of Operations & Production Management,* 20(6), 692–704.

Bougnol, M.L., Dulá, J.H., & Rouse, P. (2012). Interior point methods in DEA to determine non-zero multiplier weights. *Computers & Operations Research*, 39(3), 698–708.

Bourne, M., Kennerley, M., & Franco-Santos, M. (2004). Managing through measures: a study of impact on performance. *Journal of Manufacturing Technology Management*, 16(4), 373–395.

Bourne, M., Mills, J., Wilcox, M., Neely, A., & Platts, K. (2000). Designing, implementing and updating performance measurement systems. *International Journal of Operations & Production Management*, 20(7), 754–771.

Bovaird, T. (2008). Political economy perspectives on performance measurement. In R. Thorpe, & J. Holloway (Eds.), *Performance management: multidisciplinary perspectives* (pp. 184–199). London: Palgrave Macmillan.

Bredrup, H. (1995a). Background for performance management. In A. Rolstadas (Ed.), *Performance management – a business process benchmarking approach* (pp. 61–87). London et al.: Chapman & Hall.

Bredrup, H. (1995b). Performance measurement. In A. Rolstadas (Ed.), *Performance management – a business process benchmarking approach* (pp. 169–190). London et al.: Chapman & Hall.

Bredrup, H. (1995c). Performance evaluation. In A. Rolstadas (Ed.), *Performance management – a business process benchmarking approach* (pp. 191–198). London et al.: Chapman & Hall.

Bredrup, H., & Bredrup, R. (1995). Performance planning to ensure business achievements. In A. Rolstadas (Ed.), *Performance management – a business process benchmarking approach* (pp. 103–134). London et al.: Chapman & Hall.

Brignall, S. (2008). An accounting and finance perspective on performance measurement and management. In R. Thorpe, & J. Holloway (Eds.), *Performance management: multidisciplinary perspectives* (pp. 160–171). London: Palgrave Macmillan.

Brissimis, S.N., & Zervopoulos, P.D. (2012). Developing a step-by-step effectiveness assessment model for customer-oriented service organizations. *European Journal of Operational Research*, 223(1), 226–233.

Bruns, W.J., & McKinnon, S.M. (1992). Performance evaluation and managers' descriptions of tasks and activities. In W.J. Bruns (Ed.), *Performance measurement, evaluation, and incentives* (pp. 17–36). Boston, Massachusetts: Harvard Business School Press.

Byrnes, P., & Freeman, M. (1999). Using DEA measures of efficiency and effectiveness in contractor performance fund allocation. *Public Productivity & Management Review*, 23(2), 210–224.

Cameron, K.S. (1980). Critical questions in assessing organizational effectiveness. *Organizational Dynamics,* 9(2), 66–80.

Cameron, K.S. (1986). Effectiveness as paradox: consensus and conflict in conceptions of organizational effectiveness. *Management Science,* 32(5), 539–553.

Cameron, K.S., & Whetten, D.A. (1983a). Organizational effectiveness: one model or several? In K.S. Cameron, & D.A. Whetten (Eds.), *Organizational effectiveness; a comparison of multiple models* (pp. 1–24). New York et al.: Academic Press, Inc.

Cameron, K.S., & Whetten, D.A. (1983b). Some conclusions about organizational effectiveness. In K.S. Cameron, & D.A. Whetten (Eds.), *Organizational effectiveness; a comparison of multiple models* (pp. 261–277). New York et al.: Academic Press, Inc.

Campbell, J.P. (1979). On the nature of organizational effectiveness. In P.S. Goodman, J.M. Pennings, & Associates (Eds.), *New perspectives on organizational effectiveness* (pp. 13–55). San Francisco et al.: Jossey-Bass Publishers.

Chang, P.-L., Hwang, S.-N., & Cheng, W.-Y. (1995). Using data envelopment analysis to measure the achievement and change of regional development in Taiwan. *Journal of Environmental Management,* 43(1), 49–66.

Charnes, A.,Cooper, W.W., Lewin, A.Y., & Seiford, L.M. (1994). *Data envelopment analysis: theory, methodology and applications.* Dordrecht: Kluwer Academic Publishers.

Charnes, A., Cooper, W.W., & Rhodes, E.L. (1978). Measuring the efficiency of decision making units. *European Journal of Operational Research,* 2(6), 429–444.

Chelladurai, P. (1987). Multidimensionality and multiple perspectives of organizational effectiveness. *Journal of Sport Management,* 1(1), 37–47.

Chen, F.-H., Hsu, T.-S., & Tzeng, G.-H. (2011). A balanced scorecard approach to establish a performance evaluation and relationship model for hot spring hotels based on a hybrid MCDM model combining DEMATEL and ANP. *International Journal of Hospitality Management,* 30(4), 908–932.

Chiou, Y.-C., & Chen, Y.-H. (2006). Route-based performance evaluation of Taiwanese domestic airlines using data envelopment analysis. *Transportation Research Part E,* 42(2), 116–127.

Chiou, Y.-C., Lan, L.W., & Yen, T.-H. (2007). Integrated data envelopment analysis models for measuring transport efficiency and effectiveness. *Journal of Eastern Asia Society for Transportation Studies,* 7, 427–440.

Chiou, Y.-C., Lan, L.W., & Yen, B.T.H. (2010). A joint measurement of efficiency and effectiveness for non-storable commodities: integrated data envelopment analysis approaches. *European Journal of Operational Research,* 201(2), 477–489.

Chow, C.W., & Van der Stede, W.A. (2006). The use and usefulness of nonfinancial performance measures. *Management Accounting Quarterly,* 7(3), 1–8.

Clark, B.H. (2000). Managerial perceptions of marketing performance: efficiency, adaptability, effectiveness, and satisfaction. *Journal of Strategic Marketing,* 8(1), 3–25.

Connolly, T., Conlon, E.J., & Deutsch, S.J. (1980). Organizational effectiveness: a multiple-constituency approach. *Academy of Management Review,* 5(2), 211–217.

Cook, T.J., Vansant, J., Stewart, L., & Adrian, J. (1995). Performance measurement: lessons learned for development management. *World Development,* 23(8), 1303–1315.

Cook, W.D., & Seiford, L.M. (2009). Data envelopment analysis (DEA): thirty years on. *European Journal of Operational Research,* 192(1), 1–17.

Cook, W.D., Tone, K., & Zhu, J. (2014). Data envelopment analysis: prior to choosing a model. *Omega,* 44, 1–4.

Cook, W.D., & Zhu, J. (2007). Classifying inputs and outputs in data envelopment analysis. *European Journal of Operational Research,* 180(2), 692–699.

Cooper, W.W. (2014). Origin and development of data envelopment analysis: challenges and opportunities. *Data Envelopment Analysis Journal,* 1(1), 3–10.

Cooper, W.W., Ruiz, J.L., & Sirvent, I. (2007a). Choosing weights from alternative optimal solutions of dual multiplier models in DEA. *European Journal of Operational Research,* 180(1), 443–458.

Cooper, W.W., Ruiz, J.L., & Sirvent, I. (2009). Selecting non-zero weights to evaluate effectiveness of basketball players with DEA. *European Journal of Operational Research,* 195(2), 563–574.

Cooper, W.W., Seiford, L.M., & Tone, K. (2007b). *Data envelopment analysis: a comprehensive text with models, applications, references and DEA-Solver Software.* New York, NY: Springer Science+Business Media, LLC.

Cotte Poveda, A. (2010). Estimating effectiveness of the control of violence and socioeconomic development in Colombia: an application of dynamic data envelopment analysis and data panel approach. *Social Indicators Research,* 105(3), 343–366.

Cummings, L.L. (1983). Organizational effectiveness and organizational behavior: a critical perspective. In K.S. Cameron, & D.A. Whetten (Eds.), *Organizational effectiveness; a comparison of multiple models* (pp. 187–203). New York et al.: Academic Press, Inc.

Dimitrov, S., & Sutton, W. (2010). Promoting symmetric weight selection in data envelopment analysis: a penalty function approach. *European Journal of Operational Research,* 200(1), 281–288.

Drucker, P.F. (1963). Managing for business effectiveness. *Harvard Business Review,* May–June, 53–60.

Dyckhoff, H. (1992). *Betriebliche Produktion.* Berlin/Heidelberg: Springer.

Dyckhoff, H. (2006). *Produktionstheorie.* Berlin/Heidelberg: Springer.

Dyckhoff, H., & Ahn, H. (2010). Verallgemeinerte DEA-Modelle zur Performanceanalyse. *Zeitschrift für Betriebswirtschaft,* 80(12), 1249–1276.

Dyckhoff, H., & Allen, K. (2001). Measuring ecological efficiency with data envelopment analysis (DEA). *European Journal of Operational Research,* 132(2), 312–325.

Dyckhoff, H., Clermont, M., Dirksen, A., & Mbock, E. (2013). Measuring balanced effectiveness and efficiency of German business schools' research performance. *Journal of Business Economics,* Special issues vol. 3/2013, 39–60.

Dyckhoff, H., Mbock, E., & Gutgesell, S. (2015). Distance-based measures of specialization and balance in multi-criteria: a DEA-integrated method. *Journal of Multi-Criteria Decision Analysis,* 22(3/4), 197–212.

Dyson, R.G., Allen, R., Camanho, A.S., Podinovski, V.V., Sarrico, C.S., & Shale, E.A. (2001). Pitfalls and protocols in DEA. *European Journal of Operational Research,* 132(2), 245–259.

Dyson, R.G., & Thanassoulis, E. (1988). Reducing weight flexibility in data envelopment analysis. *Journal of the Operational Research Society,* 39(6), 563–576.

Ebnerasoul, S.A., Yavarian, H., & Azodi, M.A. (2009). Performance evaluation of organizations: an integrated data envelopment analysis and balanced scorecard approach. *International Journal of Business and Management,* 4(4), 42–48.

Eisenführ, F., Weber, M., & Langer, T. (2010). *Rational decision making.* Berlin et al.: Springer.

Emrouznejad, A., Parker, B.R., & Tavares, G. (2008). Evaluation of research in efficiency and productivity: a survey and analysis of the first 30 years of scholarly literature in DEA. *Socio-Economic Planning Sciences,* 42(3), 151–157.

Emrouznejad, A., & De Witte, K. (2010). COOPER-framework: a unified process for non-parametric projects. *European Journal of Operational Research,* 207(3), 1573–1586.

Epstein, M.K., & Henderson, J.C. (1989). Data envelopment analysis for managerial control and diagnosis. *Decision Sciences,* 20(1), 90–119.

Estellita Lins, M.P., Moreira da Silva, A.C., & Lovell, C.A.K. (2007). Avoiding infeasibility in DEA models with weight restrictions. *European Journal of Operational Research,* 181(2), 956–966.

Etzioni, A. (1960). Two approaches to organizational analysis: a critique and a suggestion. *Administrative Science Quarterly*, 5(2), 257–278.

Farrell, M.J. (1957). The measurement of productive efficiency. *Journal of Royal Statistical Society,* 120(3), 253–290.

Farzipoor Saen, R. (2010). Restricting weights in supplier selection decisions in the presence of dual-role factors. *Applied Mathematical Modelling,* 34(10), 2820–2830.

Fethi, M.D., & Pasiouras, F. (2010). Assessing bank efficiency and performance with operational research and artificial intelligence techniques: a survey. *European Journal of Operational Research*, 204(2), 189–198.

Folan, P., & Browne, J. (2005). A review of performance measurement: towards performance management. *Computers in Industry,* 56(7), 663–680.

Folan, P., Browne, J., & Jagdev, H. (2007). Performance: its meaning and content for today's business research. *Computers in Industry,* 58(7), 605–620.

Forbes, D.P. (1998). Measuring the unmeasurable: empirical studies of nonprofit organizational effectiveness from 1977 to 1997. *Nonprofit and Voluntary Sector Quarterly,* 27(2), 183–202.

Franco-Santos, M., Kennerley, M., Micheli, P., Martinez, V., Mason, S., Marr, B., Gray, D., & Neely, A. (2007). Towards a definition of a business performance measurement system. *International Journal of Operations & Production Management,* 27(8), 784–801.

García-Sánchez, I.M. (2007a). Efficiency and effectiveness of Spanish football teams: a three-stage-DEA approach. *Central European Journal of Operations Research,* 15(1), 21–45.

García-Sánchez, I.M. (2007b). Evaluating the effectiveness of the Spanish police force through data envelopment analysis. *European Journal of Law and Economics,* 23(1), 43–57.

García-Sánchez, I.M., Rodríguez-Domínguez, L., & Parra Domínguez, J. (2013). Evaluation of the efficacy and effectiveness of the Spanish security forces. *European Journal of Law and Economics,* 36(1), 57–75.

Gelade, G., & Gilbert, P. (2003). Work climate and organizational effectiveness: the application of data envelopment analysis in organizational research. *Organizational Research Methods,* 6(4), 482–501.

Ghalayini, A.M., & Noble, J.S. (1996). The changing basis of performance measurement. *International Journal of Operations & Production Management,* 16(8), 63–80.

Golany, B. (1988). An interactive MOLP procedure for the extension of DEA to effectiveness analysis. *Journal of the Operational Research Society,* 39(8), 725–734.

Golany, B., Learner, D.B., Phillips, F.Y., & Rousseau, J.J. (1990). Managing service productivity: the data envelopment analysis perspective. *Computers, Environment and Urban Systems,* 14(2), 89–102.

Golany, B., Phillips, F.Y., & Rousseau, J.J. (1993). Models for improved effectiveness based on DEA efficiency results. *IIE Transactions,* 25(6), 2–10.

Golany, B., & Roll, Y. (1989). An application procedure for DEA. *Omega,* 17(3), 237–250.

Golany, B., & Tamir, E. (1995). Evaluating efficiency – effectiveness – equality trade-offs: a data envelopment analysis approach. *Management Science,* 41(7), 1172–1184.

Goodman, P.S., & Pennings, J.M. (1979). Perspectives and issues: an introduction. In P.S. Goodman, J.M. Pennings, & Associates (Eds.), *New perspectives on organizational effectiveness* (pp. 1–12). San Francisco et al.: Jossey-Bass Publishers.

Grafton, J., Lillis, A.M., & Widener, S.K. (2010). The role of performance measurement and evaluation in building organizational capabilities and performance. *Accounting, Organizations and Society,* 35(7), 689–706.

Hamilton, S., & Chervany, N.L. (1981a). Evaluating information system effectiveness – part I: comparing evaluation approaches. *MIS Quarterly,* 5(3), 55–69.

Hamilton, S., & Chervany, N.L. (1981b). Evaluating information system effectiveness – part II: comparing evaluator viewpoints. *MIS Quarterly,* 5(3), 79–86.

Hannan, M.T., & Freeman, J. (1979). Obstacles to comparative studies. In P.S. Goodman, J.M. Pennings, & Associates (Eds.), *New perspectives on organizational effectiveness* (pp. 106–131). San Francisco et al.: Jossey-Bass Publishers.

Henri, J.-F. (2004). Performance measurement and organizational effectiveness: bridging the gap. *Managerial Finance,* 30(6), 93–123.

Hinojosa, M.A., & Mármol, A.M. (2011). Axial solutions for multiple objective linear problems. An application to target setting in DEA models with preferences. *Omega,* 39(2), 159–167.

Ho, C.-T., & Zhu, D.-S. (2004). Performance measurement of Taiwan's commercial banks. *International Journal of Productivity and Performance Management,* 53(5), 425–434.

Hsieh, L.-F., & Lin, L.-H. (2010). A performance evaluation model for international tourist hotels in Taiwan: an application of the relational network DEA. *International Journal of Hospitality Management,* 29(1), 14–24.

Huang, C.W., Ho, F.N., & Chiu, Y.H. (2014). Measurement of tourist hotels' productive efficiency, occupancy, and catering service effectiveness using a modified two-stage DEA model in Taiwan. *Omega,* 48, 49–59.

Ittner, C.D., & Larcker, D.F. (1998). Innovations in performance measurement: trends and research implications. *Journal of Management Accounting Research,* 10, 205–238.

Ittner, C.D., & Larcker, D.F. (2003). Coming up short on nonfinancial performance measurement. *Harvard Business Review,* 81(11), 88–95.

Ittner, C.D., Larcker, D.F., & Meyer, M.W. (2003a). Subjectivity and the weighting of performance measures: evidence from a balanced scorecard. *The Accounting Review,* 78(3), 725–758.

Ittner, C.D., Larcker, D.F., & Randall, T. (2003b). Performance implications of strategic performance measurement in financial services firms. *Accounting, Organizations and Society,* 28(7/8), 715–741.

Jain, S., Triantis, K.P., & Liu, S. (2011). Manufacturing performance measurement and target setting: a data envelopment analysis approach. *European Journal of Operational Research,* 214(3), 616–626.

Kao, C., Chen, L.-H., Wang, T.-Y., Kuo, S., & Horng, S.-D. (1995). Productivity improvement: efficiency approach vs. effectiveness approach. *Omega,* 23(2), 197–204.

Kaplan, R.S., & Norton, D.P. (1992). The balanced scorecard – measures that drive performance. *Harvard Business Review,* 70(1), 71–79.

Kaplan, R.S., & Norton, D.P. (1996). *The Balanced Scorecard: Translating strategy into action*. Cambridge, MA: Harvard Business Press.

Kaplan, R.S., & Norton, D.P. (2003). Keeping your balance with customers. Available at Harvard Business School: http://hbswk.hbs.edu/item/3588.html (retrieved: 12.10.2014).

Karkazis, J., & Thanassoulis, E. (1998). Assessing the effectiveness of regional development policies in Northern Greece using data envelopment analysis. *Socio-Economic Planning Sciences,* 32(2), 123–137.

Karlaftis, M.G. (2004). A DEA approach for evaluating the efficiency and effectiveness of urban transit systems. *European Journal of Operational Research,* 152(2), 354–364.

Kasie, F.M., & Belay, A.M. (2013). The impact of multi-criteria performance measurement on business performance improvement. *Journal of Industrial Engineering and Management,* 6(2), 595–625.

Keagan, D.P., Eiler, R.G., & Jones, C.R. (1989). Are your performance measures obsolete? *Management Accounting,* 70(12), 45–50.

Keeney, R.L. (1992). *Value-focused thinking: a path to creative decisionmaking.* Cambridge, MA: Harvard University Press.

Keeney, R.L., & Raiffa, H. (1993). *Decisions with multiple objectives: preferences and values tradeoffs.* Cambridge: Cambridge University Press.

Keh, H.T., Chu, S., & Xu, J. (2006). Efficiency, effectiveness and productivity of marketing in services. *European Journal of Operational Research,* 170(1), 265–276.

Khalili, M., Camanho, A.S., Portela, M.C.A.S., & Alirezaee, M.R. (2010). The measurement of relative efficiency using data envelopment analysis with assurance regions that link inputs and outputs. *European Journal of Operational Research,* 203(3), 761–770.

Klimberg, R., & Puddecombe, M. (1999). A multiple objective approach to data envelopment analysis. *Advances in Mathematical Programming and Financial Planning,* 5, 201–231.

Kumar, S., & Gulati, R. (2010). Measuring efficiency, effectiveness and performance of Indian public sector banks. *International Journal of Productivity and Performance Management,* 59(1), 51–74.

Lan, L.W., & Lin, E.T.J. (2003). Technical efficiency and service effectiveness for railways industry: DEA approaches. *Journal of the Eastern Asia Society for Transportation Studies,* 5, 2932–2947.

Lebas, M.J. (1995). Performance measurement and performance management. *International Journal of Production Economics,* 41(1), 23–35.

Lebas, M.J., & Euske, K. (2007). A conceptual and operational delineation of performance. In A. Neely (Ed.), *Business Performance Measurement: unifying theories and integrating practice* (pp. 125–139). Cambridge: Cambridge University Press.

Lee, D. (2006). Toward a way to enhance organizational effectiveness in the defense sector: associating CVF with DEA. *The Korean Journal of Defense Analysis,* 18(1), 135–159.

Lee, D., & Brower, R.S. (2006). Pushing the envelope on organizational effectiveness: combining an old framework and a sharp tool. *Public Performance & Management Review,* 30(2), 155–178.

Lengacher, D., & Cammarata, C. (2012). A two-phase data envelopment analysis model for portfolio selection. *Advances in Decision Sciences,* Vol. 2012, 1–9.

Lewin, A.Y., & Minton, J.W. (1986). Determining organizational effectiveness: another look, and an agenda for research. *Management Science,* 32(5), 514–538.

Little, J.D.C. (2004). Models and managers: the concept of a decision calculus. *Management Science,* 50(12), 1841–1853. Originally appeared (1970) in *Management Science,* 16(8), 466–485.

Liu, J.S., Lu, L.Y.Y., Lu, W.M., & Lin, B.J.Y. (2013a). Data envelopment analysis 1978–2010: a citation-based literature survey. *Omega,* 41(1), 3–15.

Liu, J.S., Lu, L.Y.Y., Lu, W.M., & Lin, B.J.Y. (2013b). A survey of DEA applications. *Omega,* 41(5), 893–902.

Liu, W.B., Zhang, D.Q., Meng, W., Li, X.X., & Xu, F. (2011). A study of DEA models without explicit inputs. *Omega,* 39(5), 472–480.

Lohman, C., Fortuin, L., & Wouters, M. (2004). Designing a performance measurement system: a case study. *European Journal of Operational Research,* 156(2), 267–286.

Lovell, C.A.K., & Pastor, J.T. (1999). Radial DEA models without inputs or without outputs. *European Journal of Operational Research,* 118(1), 46–51.

Lu, W.-M., & Hung, S.-W. (2011). Exploring the efficiency and effectiveness in global e-retailing companies. *Computers & Operations Research,* 38(9), 1351–1360.

Lynch, R.L., & Cross, K.F. (1991). *Measure up: the essential guide to measuring business performance.* London: Mandarin.

Majumder, M. (2015). Multi criteria decision making. In M. Majumder (Ed.), *Impact of urbanization on water shortage in face of climatic aberrations* (pp. 35–47). Singapore: Springer.

Marques, G., Gourc, D., & Lauras, M. (2011). Multi-criteria performance analysis for decision making in project management. *International Journal of Project Management,* 29(8), 1057–1069.

Marr, B., & Schiuma, G. (2003). Business performance measurement: past, present and future. *Management Decision,* 41(8), 680–687.

Melnyk, S.A., Bititci, U., Platts, K., Tobias, J., & Andersen, B. (2014). Is performance measurement and management fit for the future? *Management Accounting Research,* 25(2), 173–186.

Mendelow, A.L. (1983). Setting corporate goals and measuring organizational effectiveness – a practical approach. *Long Range Planning,* 16(1), 70–76.

Mensah, Y.M., Lam, K.C.K., & Werner, R.H. (2008). An approach to evaluating relative effectiveness in non-profit institutions. *Journal of Public Budgeting, Accounting & Financial Management,* 20(3), 324–354.

Meyer, M.W. (2002). *Rethinking performance measurement: beyond the balanced scorecard.* Cambridge: Cambridge University Press.

Micheli, P., & Manzoni, J.-F. (2010). Strategic performance measurement: benefits, limitations and paradoxes. *Long Range Planning*, 43(4), 465–476.

Mouzas, S. (2006). Efficiency versus effectiveness in business networks. *Journal of Business Research*, 59(10–11), 1124–1132.

Neely, A. (1999). The performance measurement revolution: why now and what next? *International Journal of Operations & Production Management*, 19(2), 205–228.

Neely, A. (2005). The evolution of performance measurement research: development in the last decade and a research agenda for the next. *International Journal of Operations & Production Management*, 25(12), 1264–1277.

Neely, A. (2007a). Introduction. In A. Neely (Ed.), *Business Performance Measurement: unifying theories and integrating practice* (pp. 1–5). Cambridge: Cambridge University Press.

Neely, A. (2007b). Part I: Performance measurement – functional analyses and theoretical foundations. In A. Neely (Ed.), *Business Performance Measurement: unifying theories and integrating practice* (pp. 7–9). Cambridge: Cambridge University Press.

Neely, A. (2007c). Measuring performance: the operations management perspective. In A. Neely (Ed.), *Business Performance Measurement: unifying theories and integrating practice* (pp. 64–81). Cambridge: Cambridge University Press.

Neely, A., Adams, C., & Kennerley, M. (2002). *The performance prism: the scorecard for measuring and managing business success*. London: Financial Times/Prentice Hall.

Neely, A., Gregory, M., & Platts, K. (1995). Performance measurement system design: a literature review and research agenda. *International Journal of Operations & Production Management*, 15(4), 80–116.

Neely, A., Kennerly, M., & Adams, C. (2007). Performance measurement frameworks: a review. In A. Neely (Ed.), *Business Performance Measurement: unifying theories and integrating practice* (pp. 143–162). Cambridge: Cambridge University Press.

Nudurupati, S.S., Bititci, U.S., Kumar, V., & Chan, F.T.S. (2011). State of the art literature review on performance measurement. *Computers & Industrial Engineering*, 60(2), 279–290.

Otley, D. (1999). Performance management: a framework for management control systems research. *Management Accounting Research*, 10(4), 363–382.

Otley, D. (2001). Extending the boundaries of management accounting research: developing systems for performance management. *British Accounting Review*, 33(3), 243–261.

Ozbek, M.E., de la Garza, J.M., & Triantis, K. (2009). Data envelopment analysis as a decision-making tool for transportation professionals. *Journal of Transportation Engineering,* 135(11), 822–831.

Paradi, J.C., & Schaffnit, C. (2004). Commercial branch performance evaluation and results communication in a Canadian bank: a DEA application. *European Journal of Operational Research,* 156(3), 719–735.

Paradi, J.C., & Sherman, H.D. (2014). Seeking greater practitioner and managerial use of DEA for benchmarking. *Data Envelopment Analysis Journal,* 1(1), 29–55.

Pavlov, A., & Bourne, M. (2011). Explaining the effects of performance measurement on performance: an organizational routines perspective. *International Journal of Operations & Production Management,* 31(1), 101–122.

Pedraja-Chaparro, F., Salinas-Jimenez, J., & Smith, P. (1997). On the role of weight restrictions in data envelopment analysis. *Journal of Productivity Analysis,* 8(2), 215–230.

Pekuri, A., Haapasalo, H., & Herrala, M. (2011). Productivity and performance management – managerial practices in the construction industry. *International Journal of Performance Measurement,* 1, 39–58.

Pennings, J.M., & Goodman, P.S. (1979). Toward a workable framework. In P.S. Goodman, J.M. Pennings, & Associates (Eds.), *New perspectives on organizational effectiveness* (pp. 146–184). San Francisco et al.: Jossey-Bass Publishers.

Portela, M.C.A.S., & Thanassoulis, E. (2006). Zero weights and non-zero slacks: different solutions to the same problem. *Annals of Operations Research,* 145(1), 129–147.

Prieto, A.M., & Zofío, J. (2001). Evaluating effectiveness in public provision of infrastructure and equipment: the case of Spanish municipalities. *Journal of Productivity Analysis,* 15(1), 41–58.

Pun, K.F., & White, A.S. (2005). A performance measurement paradigm for integrating strategy formulation: a review of systems and frameworks. *International Journal of Management Reviews,* 7(1), 49–71.

Quinn, R.E., & Rohrbaugh, J. (1983). A spatial model of effectiveness criteria: towards a competing values approach to organizational analysis. *Management Science,* 29(3), 363–377.

Radnor, Z.J., & Barnes, D. (2007). Historical analysis of performance measurement and management in operations management. *International Journal of Productivity and Performance Management,* 56(5/6), 384–396.

Ramón, N., Ruiz, J.L., & Sirvent, I. (2010). A multiplier bound approach to assess relative efficiency in DEA without slacks. *European Journal of Operational Research,* 203(1), 261–269.

Richard, P.J., Devinney, T.M., Yip, G.S., & Johnson, J. (2009). Measuring organizational performance: towards methodological best practice. *Journal of Management,* 35(3), 718–804.

Rojas, R.R. (2000). A review of models for measuring organizational effectiveness among for-profit and nonprofit organizations. *Nonprofit Management & Leadership,* 11(1), 97–104.

Roll, Y., Cook, W.D., & Golany, B. (1991). Controlling factor weights in data envelopment analysis. *IIE Transactions,* 23(1), 2–9.

Rouse, P., & Putterill, M. (2003). An integral framework for performance measurement. *Management Decision,* 41(8), 791–805.

Rouse, P., Putterill, M., & Ryan, D. (1997). Towards a general managerial framework for performance measurement: a comprehensive highway maintenance application. *Journal of Productivity Analysis,* 8(2), 127–149.

Rousseau, S., & Rousseau, R. (1997). Data envelopment analysis as a tool for constructing scientometric indicators. *Scientometrics,* 40(1), 45–56.

Rousseau, S., & Rousseau, R. (1998). The scientific wealth of European nations: taking effectiveness into account. *Scientometrics,* 42(1), 75–87.

Rutgers, M.R., & Schreurs, P. (2006). The morality of value- and purpose-rationality: the Kantian roots of Weber's foundational distinction. *Administration & Society,* 38(4), 403–421.

Sarrico, C.S., & Dyson, R.G. (2004). Restricting virtual weights in data envelopment analysis. *European Journal of Operational Research,* 159(1), 17–34.

Schick, A.G., Gordon, L.A., & Haka, S. (1990). Information overload: a temporal approach. *Accounting Organizations and Society,* 15(3), 199–220.

Schinnar, A.P., Kamis-Gould, E., Delucia, N., & Rothbard, A.B. (1990). Organizational determinants of efficiency and effectiveness in mental health partial care programs. *Health Services Research,* 25(2), 387–420.

Schmidt, S., Bateman, I., Breinlinger-O'Reilly, J., & Smith, P. (2006). A management approach that drives actions strategically: balanced scorecard in a mental health trust case study. *International Journal of Health Care Quality Assurance.* 19(2), 119–135.

Scott, W.R. (1979). Effectiveness of organizational effectiveness studies. In P.S. Goodman, J.M. Pennings, & Associates (Eds.), *New perspectives on organizational effectiveness* (pp. 63–95). San Francisco et al.: Jossey-Bass Publishers.

Seashore, S.E. (1983). A framework for an integrated model of organizational effectiveness. In K.S. Cameron, & D.A. Whetten (Eds.), *Organizational effectiveness: a comparison of multiple models* (pp. 55–70). New York et al.: Academic Press, Inc.

Sengupta, J. (2012). *Dynamics of industry growth.* New York: Springer.

Sheth, J.N., & Sisodia, R.S. (2002). Marketing productivity: issues and analysis. *Journal of Business Research,* 55(5), 349–362.

Sheth, N., & Triantis, K. (2003). Measuring and evaluating efficiency and effectiveness using goal programming and data Envelopment analysis in a fuzzy environment. *Yugoslav Journal of Operations Research,* 13(1), 35–60.

Shutt, J.A. (2003). Balancing the health care scorecard. *Managed Care,* 12(9), 42–46.

Simonen, O., Blom, M., & Viitanen, E. (2011). Knowledge of effectiveness and its application in secondary healthcare measurement. *International Journal of Productivity and Performance Management,* 60(8), 797–812.

Sinuany-Stern, Z., & Friedman, L. (1998). DEA and the discriminant analysis of ratios for ranking units. *European Journal of Operational Research,* 111(3), 470–478.

Sleezer, C.M., Hough, J.R., & Gradous, D.B. (1998). Measurement challenges in evaluation and performance improvement. *Performance Improvement Quarterly,* 11(4), 62–75.

Smith, P.C., & Goddard, M. (2008). Performance management and operational research: a marriage made in heaven? In R. Thorpe, & J. Holloway (Eds.), *Performance management: multidisciplinary perspectives* (pp. 120–136). London: Palgrave Macmillan.

Spitzer, D.R. (2007). *Transforming performance measurement: rethinking the way we measure and drive organizational success.* New York et al.: AMACOM, American Management Association.

Sprinkle, G.B., & Williamson, M.G. (2007). Experimental research in managerial accounting. In C.S. Chapman, A.G. Hopwood, & M.D. Shields (Eds.), *Handbook of management accounting research 1* (pp. 415–444). Amsterdam et al.: Elsevier.

Steers, R.M. (1975). Problems in the measurement of organizational effectiveness. *Administrative Science Quarterly,* 20(4), 546–558.

Tangen, S. (2004). Performance measurement: from philosophy to practice. *International Journal of Productivity and Performance Management,* 53(8), 726–737.

Tangen, S. (2005a). Demystifying productivity and performance. *International Journal of Productivity and Performance Management,* 54(1), 34–46.

Tangen, S. (2005b). Analysing the requirements of performance measurement systems. *Measuring Business Excellence*, 9(4), 46–54.

Taticchi, P., & Balachandran, K.R. (2008). Forward performance measurement and management integrated frameworks. *International Journal of Accounting and Information Management*, 16(2), 140–154.

Taticchi, P., Tonelli, F., & Cagnazzo, L. (2010). Performance measurement and management: a literature review and a research agenda. *Measuring Business Excellence*, 14(1), 4–18.

Tavassoli, M., Faramarzi, G.R., & Farzipoor Saen, R. (2014). Efficiency and effectiveness in airline performance using a SBM-NDEA model in the presence of shared input. *Journal of Air Transport Management*, 34, 146–153.

Taylor, D., Mrazek, M., & Mossialos, E. (2004). Regulating pharmaceutical distribution and retail pharmacy in Europe. In E. Mossialos, M. Mrazek, & T. Walley (Eds.), *Regulating pharmaceuticals in Europe: striving for efficiency, equity and quality* (pp. 196–212). Berkshire: Open University Press.

Thanassoulis, E. (1996). Assessing the effectiveness of schools with pupils of different ability using data envelopment analysis. *Journal of the Operational Research Society*, 47(1), 84–97.

Thanassoulis, E. (2001). *Introduction to the theory and application of data envelopment analysis*. Dordrecht: Kluwer Academic Publishers.

Thanassoulis, E., & Allen, R. (1998). Simulating weights restrictions in data envelopment analysis by means of unobserved DMUs. *Management Science*, 44(4), 586–594.

Thanassoulis, E., Portela, M.C.S., & Allen, R. (2004). Incorporating value judgments in DEA. In W.W. Cooper, L.M. Seiford, & J. Zhu (Eds.), *Handbook on Data Envelopment Analysis* (pp. 99–138). Boston et al.: Kluwer Academic Publishers.

Thanassoulis, E., Portela, M.C.S., & Despić, O. (2008). Data envelopment analysis: the mathematical programming approach to efficiency analysis. In H.O. Fried, C.A.K. Lovell, & S.S. Schmidt (Eds.), *The measurement of productive efficiency and productivity growth* (pp. 251–420). New York et al.: Oxford University Press.

Thomson, W. (1889). *Popular lectures and addresses*. London: Macmillan and Co. https://archive.org/stream/popularlecturesa01kelvuoft#page/n7/mode/2up (retrieved: 20.08.2014).

Ting, C.-T., Huang, C.-W. (2011). Measuring the effectiveness of mutual learning for Taiwan's tourist hotels with the DEA-Approach. *Cornell Hospitality Quarterly*, 53(1), 65–74.

Tracy, D.L., & Chen, B. (2005). A generalized model of weight restrictions in data envelopment analysis. *Journal of the Operational Research Society,* 56(4), 390–396.

Tsolas, I.E. (2010). Modeling bank branch profitability and effectiveness by means of DEA. *International Journal of Productivity and Performance Management,* 59(5), 432–451.

Tuominen, M., Rajala, A., & Möller, K. (2004). How does adaptability drive firm innovativeness? *Journal of Business Research,* 57(5), 495–506.

van Helden, G.J., Johnsen, A., & Vakkuri, J. (2012). The life-cycle approach to performance management: implications for public management and evaluation. *Evaluation,* 18(2), 159–175.

van Helden, G.J., & Reichard, C. (2013). A meta-review of public sector performance management research. *Tékhne – Review of Applied Management Studies,* 11(1), 10–20.

Venkatraman, N., & Ramanujam, V. (1986). Measurement of business performance in strategy research: a comparison of approaches. *Academy of Management Review,* 11(4), 801–814.

Vennesland, B. (2004). Measuring rural economic development in Norway using data envelopment analysis. *Forest Policy and Economics,* 7(1), 109–119.

Verschelde, M., & Rogge, N. (2012). An environment-adjusted evaluation of citizen satisfaction with local police effectiveness: evidence from a conditional data envelopment analysis approach. *European Journal of Operational Research,* 223(1), 214–225.

Vogler, S., Arts, D., & Habl, C. (2006). *Community pharmacy in Europe: Lessons from deregulation – case studies.* Vienna: ÖBIG.

Volkerink, B., de Bas, P., & van Gorp, N. (2007). *Study of regulatory restrictions in the field of pharmacies.* Rotterdam: ECORYS Nederland BV.

Waggoner, D.B., Neely, A.D., & Kennerley, M.P. (1999). The forces that shape organizational performance measurement systems: an interdisciplinary review. *International Journal of Production Economics,* 60–61(1), 53–60.

Wagner, J.M., & Shimshak, D.G. (2007). Stepwise selection of variables in data envelopment analysis: procedures and managerial perspectives. *European Journal of Operational Research,* 180(1), 57–67.

Weber, M. (1978). *Economy and society – An outline of interpretive sociology.* Berekley, CA: University of California Press.

Wholey, J.S. (1996). Formative and summative evaluation: related issues in performance measurement. *American Journal of Evaluation,* 17(2), 145–149.

Yang, J.B., Wong, B.Y., Xu, D.L., & Stewart, T.J. (2009). Integrating DEA-oriented performance assessment and target setting using interactive MOLP methods. *European Journal of Operational Research*, 195(1), 205–222.

Yu, M.-M. (2008). Assessing the technical efficiency, service effectiveness, and technical effectiveness of the world's railways through NDEA analysis. *Transportation Research Part A*, 42(10), 1283–1294.

Yu, M.-M., & Fan, C.-K. (2009). Measuring the performance of multimode bus transit: a mixed structure network DEA model. *Transportation Research Part E*, 45(3), 501–515.

Yu, M.-M., & Lee, B.C.Y. (2009). Efficiency and effectiveness of service business: evidence from international tourist hotels in Taiwan. *Tourism Management*, 30(4), 571–580.

Yu, M.-M., & Lin, E.T.J. (2008). Efficiency and effectiveness in railway performance using a multi-activity network DEA model. *Omega*, 36(6), 1005–1017.

Zammuto, R.F. (1984). A comparison of multiple constituency models of organizational effectiveness. *Academy of Management Review*, 9(4), 606–616.

List of publications

The research presented in this dissertation has partially been published in the following publications.

[1] Ahn, H., Neumann, L., & Vazquez Novoa, N. (2012a). Measuring the relative balance of DMUs. *European Journal of Operational Research,* 221(2), 417–423.

[2] Ahn, H., Neumann, L., & Vazquez Novoa, N. (2012b). Relative balance as complementary measure to relative efficiency. In: R.D. Banker, A. Emrouznejad, A.L.M. Lopes, & M.R. de Almeida (Eds.), *Data Envelopment Analysis – Theory and Applications: Proceedings of the 10th International Conference on DEA* (pp. 265–271). August 2012, Natal, Brazil.

[3] Ahn, H., & Neumann, L. (2014a). Measuring effectiveness – a DEA approach under predetermined targets. *International Journal of Business Analytics,* 1(1), 16–28.

[4] Ahn, H., & Neumann, L. (2014b). Measuring the efficiency of pharmacy stores with respect to customer retention. In D.V. Vasilenko, G.L. Bagiev, W. Fritz, O.U. Yuldasheva, & J.N. Solovjova (Eds.), *Marketing Competencies and Business Performance: Proceedings of the International Research Symposium* (pp. 46–51). May 2014, St. Petersburg, Russia.

[5] Afsharian, M., Ahn, H., & Neumann, L. (2016). Generalized DEA: an approach for supporting input/output factor determination in DEA. *Benchmarking: An International Journal,* 23(7), 1892–1909.

www.ingramcontent.com/pod-product-compliance
Ingram Content Group UK Ltd.
Pitfield, Milton Keynes, MK11 3LW, UK
UKHW020121290926
15471UKWH00003B/200

9 783631 722169